MW00462352

Hatch Guide for the
Lower Deschutes River

Hatch Guide for the Lower Deschutes River

Jim Schollmeyer

Frank Amato

PORTLAND

Dedicated To

The secret order of the Chisel Mouth Club's
other members;
Richard Bunse, Rick Hafele, Dave Hughes
and to my wife Debbie

Acknowledgements

This book would not have been possible without the
generous help of the following people:

Rick Hafele, a fly fisherman's entomologist, for helping with
the insects. Dave Hughes, my favorite writer and fishing
partner, for tremendous help with editing and moral support. My wife Debbie, for her patience, proofreading and
desire to see this completed. Richard Bunse for his art work
and fly. Ted Leeson for his clear thinking and hours of help.
Frank Amato for sharing his "home water" and publishing
this book.

Contents

Introduction

Why is there any need to identify the aquatic insects that you see clinging to streamside foliage or hiding beneath river rocks? You came to catch trout, not bugs. Besides, didn't that fly shop down the road just sell you the killer fly, the one that's supposed to be catching all the fish right now?

Nothing is absolute in fly fishing. In all likelihood the pattern that worked yesterday, or even this morning, will not be as successful later today. Trout survive by expending the least amount of energy for their food. To do this, they feed on the insect most available at any given time, whether it is a nymph drifting on the bottom, a hatching fly or an adult coming back to the water to lay its eggs. A basic understanding of aquatic insects—their life cycles, behaviors and emergence times—allows you to select the right artificial fly and fishing technique appropriate to the time of day and the season.

The lower Deschutes River can be unkind to a fly fisher not familiar with its hatches. During the past 15 years, noted entomologist and angling author Rick Hafele and his friends have collected more than 60 different aquatic insect species on the lower Deschutes. Fortunately, you won't have to deal with most of those insects—many hatch in low numbers, others are rare in most areas of the lower river and some are not available to trout at all.

The purpose of this book is to familiarize the reader with the most important stages of the fishable hatches that occur on the lower Deschutes River: the area from Pelton Dam downstream to the confluence with the Columbia River.

How to Use This Guide

Successful fishing begins well before the first cast. Take time to assess any area you intend to fish. If trout are rising, try to identify the insect on which they are feeding. When no fish are showing, check streamside foliage for insects. If none are found, turn over a few rocks in a riffle to determine which insects are most abundant and therefore most available to the trout.

A small aquarium net purchased from a pet shop works perfectly for gathering insects off the water or snaring them from the air. Carry a container in which to save captured insects; a clear film canister works well and it won't break.

With a bug in hand, try to match it to one of the insect photographs in this guide. These photos include caddisflies, stoneflies, mayflies and midges. Compare the body color of the natural to the fly in the photographs, but be aware that slight variations are common. Then measure the length of the natural on the scale provided on this page or the back cover and compare it to the size listed below the photograph. Finally, determine if the date on which you collected the insect falls within, or at least close to, the emergence time given on the hatch chart.

After identifying the insect, read the text below the photograph, which provides a description of the insect and basic information on its habitat, behavior and the times it is most active and, therefore, most important to the fish.

On the facing page, a second photograph shows three artificial flies that can be used to imitate the natural. Dressings for the flies are given at the end of the guide. If by chance you are not carrying one of the recommended patterns, pick a fly from your box that matches it most closely in size, shape and color.

The text below each group of artificals pictured provides information about the right areas, the proper times, and the best techniques to fish the flies effectively. To avoid repeating information later in the text, this guide has an early chapter desciding water types, and another explaining fishing methods. Read these carefully. The chapter on fishing methods isn't all-inclusive; use other methods or techniques that you have faith in. The main point is to deliver the fly at the right depth, in a lifelike manner. When you accomplish this, the trout will take the imitation as if it were a natural insect.

Because some minor hatches are not covered in detail in this guide, you will eventually collect an insect that you cannot find in the photographs. When this happens, choose an imitation that resembles the natural most closely and use your best judgment as to the method for fishing the fly. Keep any unknown insects for later identification. Then use one of the reference books listed in the index to complete the identification. Learning as much as you can about the insect will help you select and fish an artificial the next time you encounter the species.

```
0   5   10  15  20  25  30        40        50        60   mm
|||||||||||||||||||||||||||||||||||||||||||||||||||||||||
```

Chapter 1

The Deschutes River

The Deschutes River is nestled in the central Oregon high-desert, east of the Cascade Mountains. It flows northward for 200 miles after leaving its headwaters in the Cascades. In its last 100 miles, the river slices sharply through basalt lava flows, forming a deep and rugged canyon that is beautifully wild. After leaving the lower canyon the Deschutes joins the Columbia River on its journey to the Pacific Ocean. This lower 100 miles of the Deschutes River is considered some of the finest trout water in the world.

In the 1930s and 40s, Wickiup and Crane Prairie reservoirs were constructed on the upper Deschutes River, to catch and store spring runoff for irrigation. This reliable supply of water has been good for farmers and ranchers, but the river has not fared as well. For 100 miles downstream of these dams, the river and its aquatic inhabitants suffer from lack of water in summer and fall. With constant demand from farming and development for every drop of water, it is doubtful that the upper Deschutes River, from Wickiup Dam down to the town of Madras, will ever be the great fishery it was before these dams were built.

In the late 1950s and early 60s Round Butte Dam, Pelton Dam and the Pelton Re-regulating Dam were all built for power generation in the middle section of the Deschutes River near the town of Madras. These dams stopped all upstream migration of steelhead and chinook salmon to their historic spawning beds in the middle river. The reservoirs also buried some of the best trout water on the Deschutes. A few miles downstream from Pelton Dam, the lower of the two power dams, the Re-regulating Dam was built to help control the water fluctuations that occur when power is generated.

The early irrigation dams on the upper Deschutes reduced 100 miles of water downstream to marginal fishing. The later power dams on the lower Deschutes helped preserve the lower 100 miles and changed it to a superb tailwater trout fishery.

The Re-regulating Dam helps stabilize water levels and temperatures for miles downstream. The plankton-rich water flowing from the reservoirs feeds many of the aquatic insects that create the major hatches on the river. Given conditions like these, any trout would love to call this rich part of the river home.

Native redside rainbow trout living in this section of the Deschutes River have weathered many difficult years of overharvest and habitat abuse. But in 1983, at the urging of Oregon Trout, the Oregon Department of Fish and Wildlife instituted a slot limit, allowing only two trout between 10 and 13 inches to be killed. Catches have improved, in both numbers and size, every year since the regulation was adopted. Before 1983 landing an 18-inch rainbow would have been considered the feat of the season. Now it's common to land a fish that size on every trip.

This book was created as a guide to identification of the major insect hatches on the lower 100 miles of the Deschutes River. This stretch has plenty of access and is the most productive part of the river in terms of both insect hatches and the trout that feed on them.

In this lower 100 miles of the river, water temperatures are normally warmer downstream than they are upstream. Most hatches come off earlier near the Columbia River and work their way upstream as the season progresses. Spring runoff normally has little effect on the lower river because of the dams, but feeder streams, mainly the Crooked and White

rivers, can cause some water clarity problems from spring runoff or mid-summer glacial melt.

Fishing the Deschutes is a wonderful experience. To learn more about this unique river, I recommend reading *Deschutes* by Dave Hughes (Frank Amato Publications).

Chapter 2

Deschutes Water Types

All trout streams are not created equal, so all parts of one stream are not alike in their capacity to rear insects and grow trout. Understanding the different water types and how insects and trout utilize them, will put more jumping trout on the end of your line.

Riffles

Riffles are the fast-food counters of the fish world. Trout have the same feelings that we have: urgency, aggression and a tendency to gulp their food on the run.

Riffles are those portions of the river where the water runs fast over a shallow bottom of gravel or cobble. The size of the stones on the bottom determines the height of the chop on the surface. Water depth ranges from a few inches to three feet. Riffles are shallower at the edges and deepen toward the middle of the river. They can be very short or run on for 50 yards or more; they may be uniformly textured or sprinkled along their length with large rocks or intermittent depressions.

The fast, shallow water of a riffle is one of the richest insect areas in any river. Trout jockey aggressively for the best feeding positions in and around them. Deschutes riffles offer lots of food, but give trout little time to decide whether an item drifting past is edible. They must take an insect or an artificial, in a hurry, which makes riffles among the easiest places to fish.

Shallow water, without cover, makes trout very nervous. They will hold anywhere in a riffle if the water is deep enough. With careful observation and a little forethought, you can read the water, locate most potential holding lies and cover most of the fish in a riffle.

Runs

Trout look upon a good run in the way we look upon a nice restaurant; the seating is comfortable, there's time to read the menu and you never feel hurried. Runs move at an even pace, from a modest speed to a fast, strong flow. They may be knee-deep or over your head. Their bottoms tend to have larger rocks than riffles and more ledges and bedrock. The surface of a run rolls along with little or no chop; it often has boils or swirls caused by larger submerged boulders.

Runs are secure places for trout. They offer protective depth, numerous holding lies and current speeds moderate enough that trout have time to inspect food as it drifts past.

Pools

Fish feeding in pools are as relaxed as we are when eating at home. We feel safe and comfortable. We have time to pick and choose from the available food and if it runs out we can go out to find more.

A pool is the deepest spot in any section of the river. Water at the head of a pool, that enters from the bottom end of a riffle or run, can be fast and choppy. Often the head of a pool has eddies off to the sides, where trout love to hold. Where the pool deepens, the water slows down and the surface flattens out. As the bottom starts to lift again toward the tailout of a pool, the water will either shallow out over a very smooth bottom or else remain deep and drop quickly over a rock ledge into another fast riffle or run.

At the head of a pool, food gets delivered from the shallower waters upstream. Trout move from the deeper water in the body of a pool into this entry area to intercept insects as they drift on the current. The deeper parts of a pool offer security; the edge eddies, tailouts and head of the pool offer feeding stations, where trout can leisurely sip the supply of insects that flow in and out of the pool. When hatches are sparse trout must move out of the pool, which is generally a poor source for aquatic insects, into richer riffles, runs or flats in order to find food.

Stealth and long leaders are necessary when fishing pools; their smooth surface allow trout to notice the slightest movement a fisherman may make.

Flats

Flats to a fish are a little like cocktail parties to us; trays of delicacies may tempt you to stay in one spot to nibble or you may move around and sample the variety.

The water flow through flats is slow enough that weedbeds can take root on bottoms where silt, sand and smaller rocks have settled. The surface of the water may look deceptively smooth, but often, nearly invisible tendrils of current will cause drag on a leader that is not cast with enough slack. Weedbeds always hold an abundance of aquatic insects and when they hatch, they attract many trout. Fish feeding on flats offer one of the greatest challenges for any fly fisher. The many conflicting currents drag your fly and the slow, shallow, clear water makes trout spooky.

When fish feed on Deschutes River flats, your fishing should not be rushed. Take it slowly; fish one trout carefully and thoroughly before moving on to cover another. It's just like mingling with good friends at a cocktail party.

Pocket and Edge Water

Have you ever sat peacefully at a sidewalk cafe and watched traffic rush by? That's how a trout must feel while resting in a pocket or piece of edge water alongside fast current. But there is a major difference: trout must dash into the fast flow to take any food swirling past their position.

Pockets are small areas of quieter water downstream from obstructions in rapids. They form behind rocks, ledges and over abrupt trenches in the bottom. Trout feel secure here and food flows by just inches away.

Edge water forms in sheltered pockets of water next to the bank alongside a riffle, run, pool or rapid. To pick out productive edge water on the Deschutes, look for holding areas with a good current that comes close to shore. Some excellent edges form along the rip-rap adjacent to railroad grades; around half-submerged, grass-covered rocks next to the shore, at current seams and beneath undercut banks with over-hanging brush.

When hatches are frequent, trout move to shorelines to feed on insects trapped and drifting in edge currents. This is a good time to explore edge water with a dry fly.

Eddies

Eddies are vacation spots for trout, where they take on a kind of laid-back attitude. You can watch them resting in Deschutes eddies just about anywhere the currents are to their liking. When a hatch starts, trout move to comfortable feeding stations at the head, tail or edge of an eddy, then leisurely choose from the many insects the river has provided.

Eddies are formed where the main current flows out and away from a bank, usually where a point of land projects out into the current. This diversion of water causes a low pressure area at the top end of the eddy. This draws water in from the main current swirling it up from the bottom end of the eddy to the top, forming a continuous, circular counter-flow to the main current. An eddy has a current that flows at a slower speed, back upstream next to shore, while the main current rushes by on its outside edge.

Deschutes eddies come in all sizes and shapes. They can be deep or shallow and have flat or seething surfaces. Trout sip insects trapped in the continuous flow. Skill and patience are required to hook these trout.

Chapter 3

Fishing Methods

Deschutes River trout make their living eating insects and the ability to match the prevailing natural insect with your fly pattern puts you one step closer to having a fish on the end of your line. But it helps to go a little further; it's also important to fish the right fly at the proper depth, imparting movement or not, depending on the actions of the natural. Put this all together in the right kind of water, at the right time of day and you're sure to feel the tug of many good trout during a day on the Deschutes.

A number of excellent books have been written about casting techniques and fishing strategies. Study and practice the lessons that they offer. The following notes on methods are meant to give you a basic background in important techniques that will help you present your fly in a manner that best imitates the behavior of the natural insects.

Upstream Dry Fly Fishing

This is the standard dry fly technique. Though you can cast straight upstream, especially on rough water, it's usually best to cast the fly quartering upstream and across the current from your position. This puts the fly over the fish, but keeps the line out its feeding lane. When a cast straightens out directly upstream, the line flashes in the air above the fish and lands on the water over its head. The fly floats back down only after the line tip and leader have already passed over the fish. A cast that quarters across, keeps leader and line well away from the trout.

The Reach Cast

The reach cast is ideal for presenting your fly to the fish first, before the leader or line have a chance to spook it. After you've mastered it, this simple cast will change your life as a fly fisher.

To execute the reach cast, simply make a normal forward cast, at a 45 to 90 degree angle across the current. When the line is just starting to straighten out in the air, quickly reach your casting arm out and tilt your rod far upstream. This pulls the line upstream, but lets the fly arrive at its original destination. The fly will land on the water downstream from your leader and fly line. Now follow the drift of the fly downstream with the tip of your rod. When it passes a point straight across from you, extend your arm and reach with your rod as far as possible downstream, following the fly to lengthen its drag-free float.

Shallow Nymph

Have you ever walked up to a stream and spooked a fish out of the shallows? Or when wading, have you turned toward shore and noticed a trout close to the bank behind you, taking insects just under the surface? Have you fished water so shallow that your nymph constantly snagged the bottom? Or have you fished a nymph over weedbeds and ended up with weeds hanging on your fly on every cast? What about those fish you see working just under the surface in the evening? When you cast to them does your fly sink too deep or not deep enough? When light is low, do you have

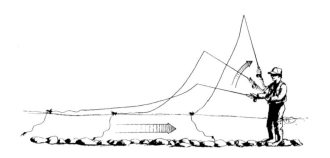

trouble telling when a trout takes your nymph? A single, simple method will help you solve all of these problems.

Put a small strike indicator on your leader, positioning it above your nymph at twice the depth you want to fish. Then fish the nymph just as you would a dry fly. Or you can go one step further and use a high-floating dry fly as your indicator. Caddis, stonefly or hopper patterns work best. You'll then have a chance at taking a fish on the dry fly as well as the nymph. Tie an 18 inch section of tippet to the bend of the dry fly hook with an improved clinch knot and tie the nymph to the end of this "dropper." You will need to slow down your casting stroke and open your loop to avoid tangling.

With this method you can follow the drift of your nymph easily by watching the indicator or the dry fly. When fishing shallow water, adjust the length of the dropper leader to keep the nymph off bottom rocks or above weedbeds. You can fish small nymphs, or even tiny dry flies, without straining your eyes. Just set the hook when the indicator pauses, sinks or if a fish strikes at the dry fly indicator.

Short-Line Nymphing

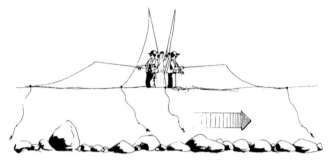

Short-line nymphing is a very intense method of fishing. It works best in Deschutes riffles, where you can get close to the fish. It requires concentration, patience and visual alertness. Like a great blue heron, the short-line nymph angler is always tense and ready to strike.

Start out with a weighted nymph or pinch small split shot onto your leader eight inches to a foot above an unweighted fly. The number of shot depends upon the depth of the water and the speed of the current. Fish with just five to 10 feet of line past the end of your rod tip. Flip the fly upstream into the riffle, then raise your casting arm and rod into the air to lift the line off the water. Follow the line tip or strike indicator downstream with the rod. Try for a drag-free drift without lifting the fly off the bottom.

You can watch the tip of the fly line to detect a strike, but a more visible, floating strike indicator helps reduce eye strain. If the line tip or strike indicator hesitates, darts or does anything else unusual, set the hook with a short flick of the rod. If it's a fish, you'll hook it. If you've momentarily caught the bottom, which should in fact happen often if you're nymphing properly, drop the rod tip and continue fishing out the drift.

Deep-Nymphing

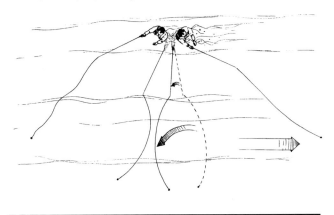

This deep-nymphing method is used on the Deschutes River more than most people would like, because it produces fish when no trout are working on the surface. It is always effective in riffles and runs. It also works well between hatches throughout the year and in winter when hatches are sparse.

Use a strike indicator on your leader for this method, set

it to twice the depth of the water you intend to fish. Select a weighted nymph or add weight to the leader 10 to 18 inches above the fly. You want your fly to get down quickly and bounce along the bottom every few feet. If it's too heavy and hangs up often, reduce the weight. If it never touches bottom, add weight.

Casting with this much weight is no joy. Use a slow casting stroke. Start out with a short line, casting a wide loop 45 degrees upstream and across. Lower your rod when the fly hits the water. Don't pull the fly off its line of drift by lifting the rod too soon. As the indicator floats back toward you, start lifting your rod tip and follow the indicator down through the drift. Lift the rod just high enough to take up the slack line, but don't pull on the indicator. As the indicator drifts past you, mend the line upstream from the indicator. This will extend your drift and keep the fly from dragging. After mending, start dropping your rod tip and extending your arm out and downstream, following the indicator, to squeeze the last inch of drift out of each cast.

At the end of the drift the fly will start to rise up from the bottom. If the current is strong enough it will lift your weighted fly almost to the surface, ready for the next cast. If you start the next cast before the fly lifts up or if the current is slow and fails to raise the fly from the bottom, you will need to lift it with a roll cast or strip in line and start again. Your rod wasn't designed to cast a weighted fly from the bottom of the river.

Trout might take the fly at any point in the drift, causing the indicator to move. During the upstream half of the drift you must set the hook quickly, with a sharp flick of the rod tip or you will miss the fish. If it's the bottom, rather than a fish, drop the rod tip and continue fishing out the drift. On the downstream part of the drift, the current will pin the fly in the trout's mouth for a moment. Set the hook with a firm, but not swift, lifting of the rod tip.

Droppers and Tailing Flies

Fishing is a lot like gambling: you place your bet on your skill and your choice of fly. It is hard to bluff a fish, even though it has a brain smaller than the tip of your little finger, and like most gamblers we are losers far too often. If you want to double your odds, add another fly to your tippet. With two flies you have twice the chance to fool the fish and cash in on a great fight to boot. You can legally fish three flies in Oregon, but two are far easier to keep from tangling and your odds don't improve much when you add a third fly, anyway.

Dropper flies are usually smaller nymphs or unweighted nymphs, tied to the leader on the long tag end of a blood knot, a foot or more above the larger point fly. Tailing flies, fixed at the end of 15 to 30 inches of tippet, are clinch-knotted to the hook bend of the point fly, which can be either a dry fly or nymph. When using a dry as the point fly, it must be a good floater. The trailing fly should never be so heavy that it pulls the dry fly under or you'll have no way of knowing when a trout takes the trailer.

Chapter 4

Caddisflies

Caddisflies go through complete metamorphosis, which means they have four different stages of development: egg, larva, pupa and adult. Most species take one year to complete the four stages.

Most of the caddisfly's life is spent in the larval stage. Caddis larvae that live in cases are called cased-caddis. They construct these cases from a wide variety of materials and carry them wherever they go. Each time the cased-caddis larva molts, casting its old skin to increase in size, the larva must either add to its case or else discard it and build a new one to make room for growth.

Caddisfly larvae that live without a moveable case are called free-living caddis. Some build permanent protective shelters, while others roam the river bed free from any case or shelter at all.

Prior to pupation, cased-caddis larvae anchor their cases to the substrate and close the opening. Free-living caddis larvae build shelters to enclose themselves for pupation. The transformation from larvae to mature pupae generally takes two to four weeks. The mature caddis pupae then cut out of their pupal cases and swim to the surface where they emerge into the adult stage (some pupae are helped to the surface by gas trapped under their pupal skins). In the surface film, the pupal shuck splits open and a fully developed adult flies off.

Adult caddis mate on streamside foliage. Females fly back to the water to deposit their eggs. Some oviposit their eggs by touching down on the water. Others dive into the water, swim to the bottom to lay their eggs on the substrate and then drift or swim back to the surface.

Green Rock Worm

Family: Rhyacophilidae
Genus: *Rhyacophila*
J F <u>M</u> <u>A</u> <u>M</u> J <u>J</u> <u>A</u> S O N D

Size: 7-12 mm
Body Color: Green
Notice: First thoracic
segment is shielded

This free-living caddis larva is not protected or concealed by a case. It is predacious, crawling among bottom rocks in cool, well-oxygenated Deschutes riffles. Hunting for its meals exposes it to strong currents. During high water flows of spring, many green rock worm larvae lose their grip on the bottom and are swept away. If luck is with them they catch the bottom again. If it's not, they make fat snacks for hungry trout. Hatches are heaviest in the first 50 miles below Pelton Dam. Warmer water conditions in the lower river, closer to the Columbia, reduce the numbers of this caddis in the last 50 miles.

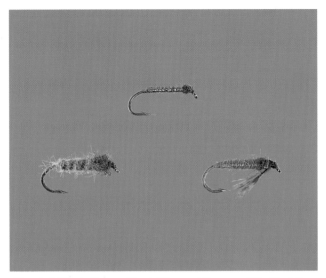

**Krystal Flash Green Rock Worm, Rhyacophila Caddis,
LaFontaine's Caddis Larva
Hook Size: 12-16**

Green rock worms reside in riffles, so spend your time fishing their imitations in riffles or in runs below riffles. Use the short-line nymphing method, with a strike indicator. The fly must be close to the bottom. If needed, add weight to the leader above the tippet knot. Cast upstream into the riffle and try to make the drift drag-free. Extend subsequent casts to cover water farther out. Takes can be very gentle. If the indicator makes any movement other than its normal drift, set the hook with a short twitch.

The shallow-water nymphing method is a great way to catch large trout that hug the edges and shallows of a riffle. Use a Krystal Flash Green Rock Worm trailer tied 20 inches below a high-floating dry fly.

Green Sedge Pupa

Family: Rhyacophilidae **Size: 8-12 mm**
Genus: *Rhyacophila* **Body Color: Green to brown**
J F M A <u>M</u> <u>J</u> <u>J</u> <u>A</u> S O N D

Free-living green rock worm larvae construct shelters of gravel and pupation takes place in riffles. When mature, the pupae leave these shelters and swim for the surface. Pupae swim well, but swift currents push them downstream as they approach the surface to emerge. Trout feed on them all the way up. Emergence occurs in the afternoon, in or downstream from riffles. You may see fish boil or even break the surface with aggressive takes. These splashy rises might make you think the fish are taking adult caddis off the surface, but most trout are rising for pupae just below the surface.

**Emergent Sparkle Pupa, Krystal Flash Pupa, Partridge
and Green Soft-Hackle
Hook Size: 12-14**

You might not notice fish working, but if you see caddis flying over the water above any Deschutes riffle in the afternoon, try a green or tan caddis pupa imitation. Fish upstream with a weighted fly, using the deep-nymphing method. At the end of every drift gently twitch the line a few times, as the fly lifts off the bottom (to imitate a swimming pupa). Often, a take to a dead-drifting fly will hardly move the indicator or line tip, but on the last part of the drift, trout hit hard and fast. If you are heavy-handed on the strike, the fly is lost. If trout work close to the surface, tie on an Emergent Caddis Pupa and use the shallow-nymph method. Fish drag-free upstream from the fish. If no fish take, then gently twitch the fly a few times on the next drift. Takes will be easy to see; this is fishing at its best.

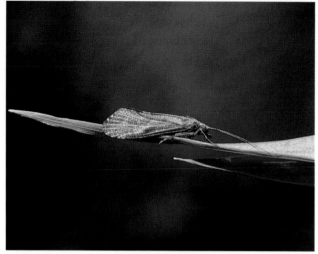

Green Sedge

Family: Rhyacophilidae   **Size: 8-10 mm**
Genus: *Rhyacophila*   **Body Color: Gray-green**
J F M A M <u>J</u> J A S O N D

Green sedge adults emerge in the afternoon, either in or downstream from riffles. During emergence, the adults don't spend much time on the water, so fish tend to feed on pupae as they swim to the surface. In late afternoon, adult females return to the water to deposit their eggs. They crawl or swim to the bottom, where they lay their eggs. When finished they drift and swim back to the surface. Fish feed on egg-laying females on the way to or from the bottom. Since emergence and egg-laying can overlap, fish might key on one phase and ignore the other, creating a puzzle for you to solve in the fading evening light.

**Deer Hair Caddis (olive), Diving Caddis (green),
Krystal Flash Diving Caddis
Hook Size: 12-14**

If you shake a streamside branch and green sedge fly out, fish the Deer Hair Caddis dry as a searching pattern. Fish it during the day on water close to over-hanging trees and grass, where the naturals may fall or be blown into the water.

Evening is the time for diving caddis patterns, fished in and downstream from riffles. Use the shallow-nymph method with a strike indicator. Fish dead-drift downstream or down and across with motion. Keep an eye on eddies downstream from riffles. Large numbers of fish often hold in these eddies to feed on adult egg-laying caddis that get swept there by the current. In eddies use the shallow nymph method with a Deer Hair Caddis as the indicator and a pupal pattern as the trailer, fished dead-drift.

Spotted Sedge Larva

Family: Hydropsychidae
Genus: *Hydropsyche*
J F M A <u>M</u> J J A S O N D

Size: 8-10 mm
Body Color: Green to tan
Notice: Gill filaments along
abdominal segments

These net-spinning caddis larvae spin spider-like nets in front of crude shelters built of leaves, sticks or stones. These retreats are attached to rocks in riffles or runs. The nets collect food from the water flowing through them. If you lift rocks from the streambed to look for Hydropsyche larvae, the net will collapse against the stone and all you will see is the shelter with the larva tucked inside. Plankton-rich waters below dams, like the lower Deschutes, support vast numbers of net spinning caddis, making Hydropsyche one of the best hatches on the river. Because these insects occur in large numbers and tend to drift downstream at sunrise and sunset, trout are always on the lookout for them.

**Dubbed Caddis Larva, Latex Caddis, Gold-Ribbed
Hare's Ear
Hook Size: 12-14**

Net-spinning caddis larvae are found in moving water and they are often dislodged in fast riffles and runs. At sunrise and sunset good numbers drift naturally, looking for locations richer in food. Early morning is a good time to fish this larval stage. Use the deep-nymph method to keep the imitation close to the bottom. Fish dead-drift though the heads of pools, riffles and runs. In the evenings, when there is usually some other insect activity going on, it is best to use a pattern like the Gold-Ribbed Hare's Ear, which can pass for a larva or pupa. Fish it dead-drift with either the deep- or shallow-nymph method. Vary the depth and add soft twitches at the end of your drift as the fly swings up.

Spotted Sedge Pupa

Family: Hydropsychidae   **Size: 8-10 mm**
Genus: *Hydropsyche*   **Body Color: Tan to green**
J F M A M J J A S O N D

Pupation of net-spinning caddis occurs in the same shelter in which the larvae tended their nets. Many Hydropsyche pupae from the Deschutes River have green underbodies with tan wings and pupal shucks. Since color can be an important factor with this hatch, carry a few different colors of pupal imitations. These pupae are good swimmers, but as they near the surface they must escape their pupal shuck to emerge. This can take time. When the pupae struggle just below the surface, they float downstream in later afternoon or evening and make easy meals for trout. Fish key in on this stage and may be seen working just under the surface, with an occasional fish breaking the surface.

**Deep Sparkle Pupa (tan), Emergent Sparkle Pupa (tan),
March Brown Spider Soft-Hackle
Hook Size: 12-14**

Use the Deep Sparkle Pupa pattern in the afternoon, well before the hatch begins. Fish it dead-drift in riffles, starting with the shallow-nymph method, then moving out to the deeper water with the deep-nymph method. Trout will be on the lookout for any early arriving pupae. Fishing activity should pick up as evening approaches and fish key in on the greater numbers of caddis pupae. When fish start working close to the surface, you'll notice weak boils from fish working deep and strong boils from fish close to the surface. Then it's time to switch to a shallow-nymph method, using a pupa emerger or soft-hackle pattern fished dead-drift. Add soft twitches at the end of the drift. Watch eddies for trout working on pupae trapped there. These are challenging fish, but the rewards are worth the effort.

Spotted Sedge

Family: Hydropsychidae
Genus: *Hydropsyche*

J F M A M J J A S O N D

Size: 8-10 mm
Body Color: Olive green to tan
Notice: Antennae as long as body

Once out of their pupal shucks, the caddis adults emerge and fly to streamside foliage. They mate there and females wait for their eggs to mature. In late afternoon or evening, vast numbers of egg-laden females dive into the water of riffles and runs, swim to the bottom and lay their eggs under rocks. Then they struggle back to the surface, where they float spent or move feebly. The swimming, struggling and spent stages of this caddis occur simultaneously. Most of the trout in the river will be feeding selectively on one of the stages. This can be a frustrating time if you pick the wrong stage to imitate.

**Deer Hair Caddis (olive-brown), Canoe Fly,
Black Soft-Hackle
Hook Size: 12-14**

Numbers of spotted sedge adults in the streamside vegetation will tip you off that this hatch is in full swing. During the middle of the day, try a Deer Hair or Elk Hair Caddis along the edge water. During emergence it's better to fish a pupal imitation rather than a dry fly. In late afternoon, watch for fish working the flats, eddies or tailouts below riffles and runs. Carefully observe the rise form; this can tell you which stage the trout are taking. Boils indicate trout feeding on swimming or struggling caddis. Use the soft hackle fished dead-drift or with motion. If fish are breaking the surface, use the Deer Hair Caddis dry fly. If a low water or spent fly is needed in flats or eddies, use the Canoe Fly. Fish the dries dead-drift. Evening is a good time for the shallow-nymph method, using a soft hackle fished below the dry fly to imitate two different stages at once.

Little Western Weedy-Water Sedge Larva

Family: Brachycentridae
Genus: *Amiocentrus*
Species: *aspilus*

Size: 8-10 mm
Tube-Case Color: Dark
green and tan
Body Color: Bright green

J F M A M J J A S O N D

This caddis larva uses weeds to build a tube-case around its abdomen. Used as a portable shelter, the case is carried at all times and is enlarged as the larva grows. When the larva reaches maturity, it attaches its case to the substrate, seals both ends and transforms itself into a pupa. These larvae feed in moderate to fast flows with weed or moss cover. They live in most riffles, runs and flats on the Deschutes River. With the large numbers of these larvae in the river, some are always losing their hold and drifting downstream before catching the bottom again. At dawn and dusk, good numbers of the caddis larvae drift a short distance downstream to change areas. Trout don't ignore these caddis larvae; they eat them, tube-case and all.

**Herl Nymph, Drifting Cased Caddis, Peeking Caddis
Hook Size: 12-14**

Examining a few rocks from a likely riffle will confirm if this slow moving caddis larva is present. You should notice the insects moving a little as they cling to the rock. If anchored to the rock, they may be pupating or the cases could be empty and the hatch over. When good numbers of larvae are found on the rocks, you can assume that trout have had a chance to feed on the larvae and will continue taking larval imitations long after the larval stage is over. Cased larvae are more active in early morning and evening. When caught in the current, these larvae just tumble along close to the bottom. Use the short-line nymph method, beginning in close. Extend your cast to cover water farther out by using the deep-nymph method. Trout are in no hurry to grab the larvae, so takes can be subtle.

Little Western Weedy-Water Sedge Pupa

Family: Brachycentridae   **Size: 7-9 mm**
Genus: *Amiocentrus*   **Body Color: Green**
Species: *aspilus*
J F **M** A **M** J J A S O N D

The bright-green body color of this caddis pupa really stands out. Pupation takes place inside the same case that the larva wore while crawling around in riffles, runs and flats. After anchoring its case to a rock or other bottom debris, the larva seals itself inside to pupate. When mature, the pupa will cut its way out of the case and swim toward the surface as it drifts downstream. When floating in the surface film, the pupal shuck splits open and the adult caddis emerges. This process may take some time. Newly hatched caddis adults fly quickly to streamside vegetation. Pupal emergence starts in the afternoon and increases toward evening. From the moment the pupa leaves its case until it emerges, it's available to trout.

**Krystal Flash Pupa (green), Emergent Sparkle Pupa (green), CDC Caddis Emerger (green)
Hook Size: 14-16**

To detect this hatch, check a few submerged rocks for attached tube-cased caddis larvae that are pupating in the riffle, run or flat you are fishing. The hatch is in progress if cases are present and the tail ends of some of the tubes are open, indicating that pupae have emerged. In the afternoon, start fishing a pupal imitation dead-drift and close to the bottom. Use either the short-line or deep-nymph method, whichever matches the water you are fishing. The numbers of emerging caddis should increase toward evening, with the fish feeding on them at all depths. When fish feed in the surface film, it may be difficult to determine if they are taking adults or pupae. Most of the time it will be the pupae. Use a caddis emerger or pupal imitation with the shallow-nymph method, dead-drift or add a soft twitch as the fly nears fish feeding in the surface film.

Little Western Weedy-Water Sedge

Family: Brachycentridae **Size: 7-9 mm**
Genus: *Amiocentrus* **Body Color: Green**
Species: *aspilus*
J F M **A** M **J J** A S O N D

If you suspect that a caddis scurrying around on stream-side foliage could be the adult form of a tube-cased caddis, check submerged rocks for evidence of their pupa (see page 38). This caddis is present in most of the riffles, runs and flats on the lower Deschutes River. During the hatch, which occurs in the afternoon and on into evening, fish may feed selectively on either pupae or adults. Ovipositing females return to the water throughout the day and float serenely on the water's surface depositing their eggs. Trout love this caddis adult: it's like having a light snack to nibble on through-out the day.

**Elk Hair Caddis (olive), Sparkle Caddis (olive), CDC
Caddis (olive)
Hook Size: 14-16**

Emergence and egg-laying stages of this caddis don't produce large swarms, but occur sporadically through the afternoon and evening over riffles, runs and flats. Trout key on this irregular emergence, which makes it a logical hatch to imitate when exploring water. The CDC Caddis or Sparkle Caddis patterns are ideal for picky trout in eddies and flats. A high-floating Elk Hair Caddis is the answer for fishing edge waters, riffles and runs. This adult caddis floats with little or no movement, so fish its imitations dead-drift. If other types of caddis are present, watch the rise forms carefully. Surface rises indicate trout might be feeding on the tube-cased caddis adults. Sub-surface boils or splashy rises signal that fish are feeding on pupae or on a different caddis. Fish a pupal pattern or try to identify and imitate the other caddis.

Saddle-Case Caddis Larva

Family: Glossosomatidae   **Size:** 5-8 mm
Genus: *Glossosoma*   **Body Color:** Cream to tan
J F **M A M** J **J A** S O N D   **Notice:** Pebbled case

These saddle-case caddis larvae build distinctive cases, which completely cover the larvae except for small openings at both ends. Worn like a saddle, the case offers protection for the larva as it crawls over submerged rocks and feeds. Lacking gills, the larvae need the fast flowing waters of riffles for oxygen. The case is not expandable, so periodically the larva must abandon it to build a bigger one. As the larvae near maturity and grow rapidly, they change cases more often. When large numbers of insects in the same age group shed their cases, many are swept off the rocks by the current. The small size of these larvae means nothing to a trout if the supply is plentiful.

**Gold-Ribbed Hare's Ear, Cream Caddis Larva,
Latex Caddis
Hook Size: 14-18**

You will find the saddle-case caddis in riffles. Look for their stone cases attached to submerged rocks. If present, there's a good chance that the trout are already familiar with the small larvae as a food source, since large numbers of them become available when they leave their old cases to build new ones. Trout move into the riffles or hold just downstream from them to feed on larvae as they tumble along the bottom. Even if few larvae are in the drift, trout will still take a larval imitation, so it's not necessary to be there during the peak of the drift to fish this stage. The fly must be fished close to the bottom with a drag-free drift. Use the short-line nymphing method in close; then change to the deep-nymph method as you cover water farther out. Be sure to fish the water just below a riffle.

Saddle-Case Caddis Pupa

Family: Glossosomatidae   **Size: 5-8 mm**
Genus: *Glossosoma*   **Body Color: Tan to green**
J F M A M J J A S O N D

Prior to pupating, saddle-case caddis larvae often migrate to the margins of riffles. When this occurs, these miniature caddis shelters collect in immense numbers and resemble overcrowded campgrounds on the rocks. Pupation occurs in the closed cases. At maturity, the pupae cut out of their cases, swim to the surface and emerge as adults. Emergence is sporadic through the early afternoon and picks up toward evening. When other insect hatches are sparse in early spring and fall, trout feed on these small pupae, even when insect numbers are low. During summer, when other families of caddis are available, these small pupae may be overlooked by the angler, but trout will selectively feed on them when the insects are abundant.

**Hare's Ear Soft-Hackle, Little Green Caddis, Deep
Sparkle Pupa (green)
Hook Size: 14-18**

Although it is one of the smallest caddis pupae in the Deschutes, this pupa is an important food source for trout when hatches are sparse in spring and fall. If you discover their rock cases in the margins of a riffle, then observe small caddis fluttering over the water in the afternoon, take heed: these may be your only clues that trout are feeding on these small emerging pupae. Use the short-line or deep-nymph method to fish the riffles and the waters below them. Trout won't move far to take this fly, so cover the water systematically. Later in the afternoon, if you see fish feeding close to the surface, change to the shallow-nymph method and use a soft hackle fished dead-drift, with an occasional twitch. If large numbers of this small caddis are emerging, trout may selectively feed on them, ignoring the presence of other caddis.

Saddle-Case Caddis Adult

Family: Glossosmatidae   **Size: 6-8 mm**
Genus: *Glossosoma*   **Body Color: Tan to green**
J F M <u>A</u> M <u>J</u> J <u>A</u> S <u>O</u> N D

The adult saddle-case caddis looks and behaves like a smaller version of the free-living caddis. At emergence times trout are more likely to feed on the easily caught pupae than the adult, which spends little time on the water before flying off to the shore. When the egg-bearing female returns to the river, dives into the water, swims to the bottom to lay her eggs and then struggles back to the surface, trout feed heavily on this small caddis adult. Ovipositing occurs in the late afternoon and on into evening. Emergence and egg-laying stages of this caddis can happen at the same time and fish may feed selectively on either stage. Trout tend to feed more on the saddle-case caddis in the spring and fall, when other caddis hatches are meager.

**Deer Hair Caddis (brown), Starling and Herl, Krystal
Flash Diving Caddis
Hook Size: 14-18**

Look for this adult caddis on streamside foliage. If they are present in large numbers, waters downstream from riffles should produce good fishing in the afternoon when ovipositing takes place. This egg-laying caddis is an active swimmer, but after laying its eggs, the adult struggles back to the surface, where it may lay spent on the water. Use a diving caddis pattern with the shallow-nymph method. Fish it dead-drift on the top half of the drift to simulate the spent adult, and then twitch it through the bottom half of the drift to imitate the swimming adult. Emergence and egg-laying may occur at the same time. When in doubt, change the diving caddis to a soft hackle, which can imitate either the pupa or the swimming adult. The Deer Hair Caddis imitates the spent adult on the surface and may be used by itself or as the indicator for the shallow-nymph method. Eddies below riffles are great collection spots for the spent adults of this caddis.

Chapter 6

Stoneflies

Stoneflies have incomplete metamorphosis, going through three stages: egg, nymph and adult. The time spent in each stage varies with each species. Their life cycles last one year for the smaller stonefly species and two or three years for larger stoneflies, such as the golden stones and salmonflies.

Most of the stonefly's life is spent in the nymphal stage. Mature nymphs crawl out of the water, usually at evening or after dark, and the adults emerge from their nymphal shucks on streamside foliage, where they are safe from trout.

Adults mate in the foliage or on the ground. After mating the females fly out over the water and dip to the surface to release their eggs.

Giant Salmonfly Nymph

Family: Pteronarcidae **Size: 25-50 mm**
Genus: *Pteronarcys* **Body Color: Dark-brown to black**
Species: *californica* **Notice: Gill tufts on thoracic**
 abdominal segments

<u>J</u> <u>F</u> <u>M</u> <u>A</u> <u>M</u> <u>J</u> <u>J</u> <u>A</u> <u>S</u> <u>O</u> <u>N</u> <u>D</u>

The largest stonefly in North America may not be a giant in our eyes, but when you compare its one to two inch body length to the size of other insects in the river, it is indeed gigantic. The nymph lives in riffles and runs and around rocks that collect plant debris, which is food for these docile grazers. They take an average of three years to mature, so there are always nymphs of different sizes present. Their numbers can be so great that they have been likened to the huge buffalo herds of the past. As emergence draws near, the nymphs migrate toward shore in the afternoon and evening. Trout will feed on giant stonefly nymphs whenever they are available. Emergence coincides with trout spawning, making this nymph a very important food source for Deschutes trout.

**Kaufmann's Black Stone, Box Canyon Stone,
Lead-Eyed Woolly Bugger (black)
Hook Size: 4-8, 3X long**

When salmonfly nymphs migrate from the deeper water of riffles and runs toward the shallows, trout wait to intercept them. In the morning and evening, fish a lightly weighted nymph using the shallow-nymph method along the margins of riffles and runs. Then switch to the deep-nymph method as you fish to the deeper water. Keep the fly dead-drifting close to the bottom. Because of the three years it takes these insects to mature, giant stonefly nymphs are continually available to trout. A weighted stonefly nymph is an excellent pattern to fish at any time of the year. To increase your odds, add a smaller dropper or trailing fly that matches the most prevalent insect in the riffle. Through the fall and winter months, a steelhead may well lay claim to your stonefly nymph—a guaranteed surprise, especially if you're using a light tippet.

Giant Salmonfly

Family: Pteronarcidae   **Size:** 25-50 mm
Genus: *Pteronarcys*   **Body Color:** Dark-gray with
Species: *californica*   reddish-orange underside
J F M A <u>M</u> J J A S O N D

    You'll have no doubt when this hatch occurs. Adults advertise their presence by crawling all over streamside foliage and sometimes down your neck. From evening to early morning, emerging nymphs crawl out of the water. Once ashore, they cling to streamside objects; the nymphal shuck splits open, allowing the adults to emerge. Adults spend most of their time on foliage next to the river; many fall or are blown into the water when the afternoon wind arrives. After mating, the females fly out over the water, normally in the afternoon, releasing their eggs as they touch the water's surface. Adult stoneflies are not good fliers and many end up on the water during their egg-laying flights.

Bird's Stone, Improved Sofa Pillow, Clark's Stonefly
Hook Size: 4-8, 3X long

When the streamside brush is crawling with adult giant salmonflies, it's time to tie on a dry stonefly pattern. Trout take time to adjust to this large insect struggling on the water. But once accepted as edible, any adult stonefly hitting the water is in big trouble. Use dry stonefly patterns after the adults get active later in the morning, then continue into the afternoon until trout lose interest. Fish water downstream from riffles, along runs and in edge water next to banks with good cover. If trout are not interested in a dead-drift presentation, try twitching the fly during its drift. Trout will continue taking dry stonefly patterns for days after the hatch is over. These large patterns require stout 3X or 4X leaders to cast properly and to withstand the sudden, violent rise of a trout—or the surprised, aggressive hook set by the angler.

Golden Stone Nymph

Family: Perlidae
Genus: *Hesperoperla*

J F M A M J J A S O N D

Size: 22-35 mm
Body Color: Yellow-tan to light brown
Notice: Gill tufts at the base of each leg

Golden stone nymphs live in riffles and runs where they prey on smaller aquatic insects. They are very active, pursuing their quarry among rocks in fast currents. This willingness to move about means that many nymphs are swept off the bottom and drift downstream. It takes two years for these nymphs to mature, so they are always available to trout. Mature nymphs migrate to shore during the afternoon and evening and crawl out of the water to emerge as adults. This migration can be very heavy at times and the trout will follow the nymphs into very shallow water to feed on them. The golden stone emergence begins right in the middle of the giant salmonfly hatch, which causes the golden stone nymph to be overlooked by anglers, but not by trout.

**Golden Stone Nymph, Yellow Stone Nymph,
Kaufmann's Golden Stone
Hook Size: 6-8, 3X long**

Golden stone nymphs pursue their prey in riffles and runs. That's where you should look for trout feeding on nymphs that get swept off rocks. Use the short-line nymph method to fish the water in close; then change to the deep-nymph method and extend your casts to cover the water farther out. The fly should tumble along close to the bottom. If you don't make contact with the bottom, add more weight to the leader. When the nymphs migrate toward the banks prior to emergence in the afternoon and evening, trout will move into sheltered holding positions close to the edges of riffles and runs to feed on them. This is the time to use the shallow-nymph method with a lightly weighted nymph pattern, like the Golden Stone. Go slow and cover any likely holding water. Takes will be quick and hard. This is exciting fishing, and at times you will be able to spot the fish before you cast to it.

Golden Stonefly

Family: Perlidae
Genus: *Hesperoperla*
J F M A <u>M</u> <u>J</u> J A S O N D

Size: 22-35 mm
Body Color: Golden brown
Notice: Nymphal gill remnants at the base of each leg

After a mature golden stone nymph crawls out of the water onto streamside foliage, the nymph's exoskeleton splits open and the adult slowly squirms out, leaving its nymphal shuck behind. As the hatch progresses, hundreds of these shucks will litter streamside foliage. Mating adults remain in the brush next to the river, and many fall or are blown into the water by the wind. In the afternoon or evening, the golden stonefly female laden with eggs flies out over the river and drops to the water's surface for a quick release of her eggs before fluttering back into the air. Since these stoneflies are clumsy fliers, many end up on the water. Trout rush to grab any helpless adult caught on the water's surface.

**Bucktail Caddis (yellow), Stimulator (golden),
Sofa Pillow
Hook Size: 6-8, 3X long**

If you enjoy catching large, wild redsided rainbow trout on a dry fly, the golden stonefly hatch is one of the best times to do so. Emergence starts right in the middle of the giant salmon fly hatch. Trout are accustomed to feeding on these larger flies, and when the golden stonefly appears, the fish waste little time accepting this somewhat smaller insect. This hatch extends the time during which you can fish a big dry fly over trout, since golden stoneflies are available well into June. As the day warms and the wind starts to blow in the afternoon, tie on a golden stonefly pattern. With stealth and patience, fish the edge water for any trout waiting to feed on stoneflies that fall from the streamside foliage. When females begin their egg-laying flights, work the water around riffles and runs. The explosive takes surprise many anglers.

Little Yellow Stonefly Nymph

Family: Perlodidae   **Size: 7-16 mm**
Genus: *Cultus*   **Body Color: Yellow to**
                                 light brown

J F M A M J̲ J̲ A S O N D   **Notice: Absence of**
                                 external gills

    Little yellow stonefly nymphs take less than a year to mature. They live in riffles and runs, where fast currents supply food and well-oxygenated water. Nymphs of this genus are predators. Strong currents often dislodge them as they actively pursue their prey. The nymphs' greatest time of vulnerability to trout occurs during their migration to shore, prior to emergence. This migration, which takes place throughout the day, causes large numbers of these nymphs to get swept downstream. They tumble along until they catch the bottom or are captured by a trout. Nymphs that reach the banks crawl out of the water onto streamside rocks and foliage, to emerge as adults.

**Little Yellow Stone, Stonefly Creeper, Maggot
Hook Size: 10-12, 2X long**

This stage is often overlooked by anglers who are still focused on the last days of the golden stonefly hatch. Check rocks and brush along the shore for the small, empty nymphal shucks of the little yellow stoneflies. If they are present, don't miss the opportunity to fish this stage. Start with the shallow-nymph method and fish the margins along riffles and runs. Next, change to the short-line method to fish the riffles in close. Finish by using the deep-nymph method to cover the water farther out in the riffles or runs. Pay close attention to shallow riffles that drop off into deeper water. Trout will often hold in the deep water close to the edge of the drop-off to catch these drifting nymphs. Cast the nymph imitation upstream of the drop-off, so it tumbles drag-free over the edge into the deeper water. Set the hook at the slightest pause of the line or indicator.

Little Yellow Stonefly

Family: Perlodidae   **Size: 7-16 mm**
Genus: *Cultus*   **Body Color: Yellow, yellow-**
                             orange to rear of abdomen
J F M A M J J A S O N D

After leaving the water, little yellow stonefly nymphs cling to streamside rocks or foliage where they emerge as adults. This transformation normally occurs in late afternoon or evening. During the day these active little stonefly adults can often be seen crawling on foliage or fluttering out over the river. After mating, females will fly out over the river above riffles or runs, glide to the water's surface, deposit their eggs and fly back to shore. This ovipositing flight of the females usually occurs during the calm evening hours. At the peak of this hatch, it is not unusual to see very large swarms of these small female stoneflies fluttering up and down, over riffles or runs.

**Bucktail Caddis (yellow), Stimulator (yellow), Clark's
Little Yellow Stone
Hook Size: 10-12, 2X long**

As the morning chill leaves the canyon, these little stoneflies start moving, crawling on streamside foliage and flying out over the river. They are good fliers, but when the afternoon winds blow, some of these adults will end up on the water. This is a good time to fish upstream with the wind at your back. Cast a little stonefly pattern into likely holding lies along the edges of riffles and runs or under over-hanging branches along edge waters. During mating flights, the female stoneflies touch down on the water's surface in the evening to release their eggs and trout are waiting for them. A good high-floating and visible stonefly pattern fished through riffles and runs can produce memorable fishing experiences.

Chapter 7

Mayflies

Mayflies have incomplete metamorphosis, which means that they go through only three stages of development: egg, nymph and adult. Most species have a one-year life cycle.

Mayflies spend most of their lives as nymphs. The nymphs are grouped into four different categories based on their behavior and shape: swimmers, crawlers, clingers and burrowers. The lower Deschutes River has all of the categories except burrowers.

Nymph wing pads darken just before the insects reach maturity. At emergence, the mature nymphs leave the river bottom and swim or drift to the surface film. The adults emerge from the nymphs and fly off to streamside foliage.

The adult stage actually has two phases, the dun and spinner. The dun emerges from the nymph in a winged form. Duns have opaque wings, but within hours to a few days, the dun molts and becomes a spinner. With clearly veined wings in the spinner phase, the mayfly mates and lays eggs. Male spinners collect in large swarms; the female spinners fly into the swarm and mate while still in flight. After mating, the females lay their eggs by touching down on the surface film or by crawling under the surface to deposit them on the bottom.

After laying their eggs, the flying females fall to the water's surface and die, creating a spinner fall.

Blue-Winged Olive Nymph

Family: Baetidae   **Size: 3-8 mm**
Genus: *Baetis*   **Body Color: Olive-brown to**
                              dark brown

J F M A M J J A S O N D   **Notice: Three fringed tails,**
                                  with shorter middle tail

Baetis nymphs live in most of the moving water on the Deschutes River. Some of the largest concentrations occur in shallow riffles, weedy flats and runs. These small mayfly nymphs are good swimmers, moving short distances with a quick flip of their tail. They often dart from spot to spot, even in faster currents and tend to be most active at sunrise and sunset. Prior to emergence, their wingpads will darken and become more developed. Mature nymphs drift or swim from the substrate to the surface at emergence, which generally takes place in the morning (during the warmer months) and in early to mid-afternoon (during the colder months). The multiple generations of *Baetis*, along with their large numbers, make these nymphs a common food source available to trout all year.

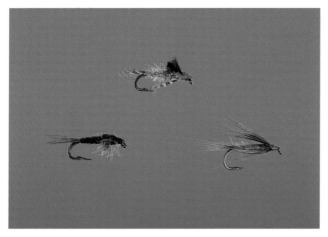

**_Baetis_ Nymph, Krystal Flash Nymph, _Baetis_ Soft-Hackle
Hook Size: 16-18, 2X long**

At sunrise and sunset, riffles and waters directly downstream from them are good areas to fish _Baetis_ nymph patterns any time of year. During winter and early spring,
when other insect activity is down, these nymphs are still
going strong. Fish their imitations all day. Use the short-line
and deep-nymph methods with a _Baetis_ nymph pattern alone
or as a dropper with a stonefly nymph pattern. During the
summer, as weed growth increases in riffles, runs and flats,
trout move into shallow water around these weedy areas in
the early morning and evening hours to feed on these
nymphs. Approach this type of water carefully, watching for
feeding fish. The shallow-nymph method, with the strike
indicator set to keep the fly above the trailing weeds, is high-
ly effective in this type of water. Fish take these mayfly
nymphs without much commotion, so tighten up on the line
at any unusual movement of the indicator.

Blue-Winged Olive Dun

Family: Baetidae	**Size: 3-8 mm**
Genus: *Baetis*	**Under Body Color: Light to dark olive**
<u>J F M A M J J A S O N D</u>	**Notice: Small, narrow hind wings**

Even though the *Baetis* emerges all year, it's not a hatch to count on. You may go days before you see a dun, but they appear often enough to attract the attention of trout. In the colder months, duns will appear in the warmest part of the day, often floating for long distances before flying off. Through the warmer months emerging duns start showing on the water's surface in cool, morning hours and fly off after a short drift. At the start of a hatch, fish often feed on the drifting nymphs, ignoring the duns. Duns fly to streamside foliage and later molt into spinners. Mating flights occur mornings and evenings. After mating, females fly to the water, deposit their eggs and die, often ending spent on the water.

**Little Olive, Hairwing Dun, Blue-Winged Olive Parachute
Hook Size: 16-18**

This hatch is like a treasure hunt: follow the clues correctly and you may end up with a reward. But don't count on it. In winter, watch the eddies below riffles, where duns will collect and attract pods of wary rainbow trout. This fishing calls for long leaders, small flies and stealth. During the warmer months fish will move into riffles, runs and flats to feed on the nymphs and emerging duns. At the start of a hatch, fish a nymph pattern, changing to a dry fly later in the hatch. Eddies offer some of the best and most difficult dry fly fishing for trout feeding on *Baetis* duns and spinners trapped in the surface film. In this situation use low water flies like parachute patterns, long fine leaders, stealth, patience and luck. They will all be necessary to hook and land the large trout that feed in Deschutes eddies. Drift the fly down to the feeding trout, so the fly, not your leader, is seen first by the trout.

Pale Morning Dun Nymph

Family: Ephemerellidae   **Size: 7-9 mm**
Genus: *Ephermerella*   **Body Color: Olive-brown**
Species: *inermis*
J F M A M J J A S O N D

Pale Morning Dun (PMD) nymphs are crawling mayflies that occupy runs and flats with moderate flows. They normally keep out of harm's way by hiding among rocks and mosses on the river bottom. If dislodged, these nymphs tumble downstream wiggling their abdomens up and down in feeble swimming motions, until they regain the bottom or have the misfortune of meeting a trout. During emergence, usually between 10 a.m. and 3 p.m., the nymphs face their greatest hazard as they swim slowly from the river's bottom to the surface, where they emerge as duns. The nymphs drift for long distances before reaching the surface, giving trout plenty of time to feed on them.

**Pale Morning Dun Nymph, Olive-brown Hare's Ear,
PMD Floating Nymph
Hook Size: 14-16**

Fish generally don't see many Pale Morning Dun nymphs until their emergence period. Once the nymphs start emerging, trout readily intercept any nymph imitation presented in a life-like manner. Emergence normally starts between 10 a.m. to 3 p.m. During this time, fish a PMD nymph pattern through runs and flats. Use the deep-nymph method to keep some distance between you and the fish, which may be spooked by wading too close. When you notice fish feeding close to the surface, change to the shallow-nymph method. Use a lightly weighted nymph to cover water one to two feet deep. When fish are rolling on the surface, they are probably taking nymphs in the surface film. Try the shallow-nymph method with an indicator three feet up the leader from a floating-nymph pattern.

Pale Morning Dun

Family: **Ephemerellidae**
Genus: *Ephermerella*
Species: *inermis*
J F M A M J̲ J̲ A̲ S O N D

Size: **7-9 mm**
Under Body Color: **Pale yellow with olive sheen**
Notice: **Distinct shape of front edge on hind wing**

Pale Morning Dun (PMD) nymphs live in runs and flats—the same areas in which the duns emerge. Not all runs and flats are productive, so when you find a good hatch, remember the location. Emergence normally occurs between 10 a.m. and 3 p.m. Nymphs reaching the surface split open and the duns emerge. Adult flies sometimes float long distances before flying off to streamside foliage. By the next day, the duns have molted into spinners, which have translucent wings and light olive-brown bodies. These spinners mate in the air upstream from their runs and flats. After mating, the females drop to the water to deposit eggs and remain there spent, floating downstream. Spinner falls occur in the afternoon.

**CDC PMD Floating Nymph/Emerger, Pale Morning
Comparadun, PMD Spinner
Hook Size: 14-16**

Any angler encountering a good PMD hatch on a flat or run should see a fair number of nice trout gently rising to the floating duns. Take time to evaluate the situation. Subsurface boils indicate trout are feeding on the nymphs. Fish rising to invisible flies may mean they are feeding on floating nymphs, emergers or spinners. The easiest to recognize are rises to the duns. Match your fly pattern to the insect stage on the water. The smooth, slow water where these duns hatch demands cautious fishing. The reach cast or downstream cast both present the fly first to feeding fish and are excellent casts for all stages of this hatch. The main objective is to pick out a feeding trout and present the fly drag-free, without the leader or line spooking the fish. This is one of the most productive mayfly hatches on the lower Deschutes, so relish the times you fish it.

Slate-Winged Olive Nymph

Family: Ephemerellidae   **Size:** 9-12 mm
Genus: *Drunella*   **Body Color:** Dark olive-brown
Species: *coloradensis*
J F M A M J J A S O N D

    Slate-Winged Olive Nymphs are crawlers. They prefer riffles and runs with cobbled bottoms. Woody debris, roots or weeds that may be present in these areas are choice habitats for the nymphs. They spend most of their time hidden on the river bottom and stay out of harm's way until emergence. In late afternoon, mature nymphs release hold of the river substrate and drift downstream as they swim slowly to the surface. After reaching the surface film, the nymphs float along until the duns emerge from the nymphal shuck. Cold water is necessary for these nymphs to survive. Warmer water temperatures in the lower fifty miles of the Deschutes River limit the range of these nymphs to the upper 50 miles.

D D D, Olive-brown Hare's Ear, Green Drake Emerger
Hook Size: 10-14

Unless you notice Slate-Winged Olive duns on the water, there is no way to tell if these mayfly nymphs are present without sampling a few rocks from the riffle or run. Look for nymphs with dark wingpads, which indicate they are close to emergence. If good numbers of the mature nymphs are found, start by using the deep-nymph method later in the afternoon. Keep the imitation close to the bottom and continue to fish deep until you notice fish feeding close to the surface. Watch the rise forms. If you see subsurface boils or rises but see nothing on the surface, they may be taking emergers. Switch to the shallow-nymph method and use an emerger pattern. Fish feeding close to the surface are usually very cautious. Don't spook them by casting too close or letting the fly line drift over them.

Slate-Winged Olive Dun

Family: Ephemerellidae
Genus: *Drunella*
Species: *coloradensis*
J F M A M J J A S O N D

Size: 9-12 mm
Underbody Color: Dark olive

One of the larger mayflies on the Deschutes River, Slate-Winged Olives are often mistaken for other mayflies from the same genus, the less common and stouter Green Drakes. As the habitats, habits and emergences are similar for both mayflies, the same techniques and fly patterns may be used to imitate them. Emergence occurs in late afternoon when the mature nymphs release their hold on bottom rocks in riffles and runs. They swim to the surface to emerge from their nymphal shucks. The struggling nymphs and newly emerged duns float for long distances on the surface; this is the most vulnerable time for these mayflies. The duns fly to streamside foliage, where they later molt and emerge as spinners. In the evenings, spinners gather in the air over riffles to mate. The females fall to the water and deposit their eggs, then lay spent, floating downstream.

**Natural Dun, Western Green Drake,
Olive Sparkle Haystack
Hook Size: 10-14**

The exquisite coloration of these large *Drunella* mayflies is a joy to see. If Slate-Winged Olives or Green Drakes are hatching in large enough numbers to bring up trout, an angler can have a memorable evening fishing to this feeding spree. The duns emerge in the evening. If no fish are seen feeding on duns, fish a nymph pattern upstream from the emerging mayflies. As the trout start feeding on the surface, watch the rise forms carefully. Trout often feed selectively on either the nymphs, emerging duns or floating duns. When in doubt about which fly to use, try an emerging-dun pattern fished dead-drift over the feeding trout. Use the high-floating Western Green Drake if the water is choppy or you have trouble spotting your fly in the low evening light. The Natural Dun rides low in the water but floats like a cork and is an excellent fly to use for trout feeding on duns or spinners in runs or the flat water below riffles.

Pale Evening Dun Nymph

Family: Heptageniidae
Genus: *Heptagenia*
Species: *solitaria*

Size: 7-10 mm
Body Color: Olive-brown
Notice: Flat head and body; three body-length tails

J F M A <u>M</u> J J A S O N D

Heptagenia are called clingers because their flat heads and bodies allow the nymphs to cling to submerged rocks even when exposed to the full force of fast currents. As these active nymphs scurry about the substrate, their tenacious grips prevent them from being swept away. Few end up in the drift and the ones that are dislodged normally return to the bottom with a few quick flips of their abdomen. Nymphs close to maturity migrate to the calm waters along the edges of riffles and runs. Adults normally emerge in the afternoon hours. Nymphs release their grip on the substrate and swim rapidly to the surface. Trout often move into this shallow, calm water to feed on the nymphs as they swim for the surface.

**Cate's Turkey, Olive-brown Pheasant Tail,
Dark Olive Soft-Hackle
Hook Size: 12-14**

If you are fishing a riffle or run early in the day and it has a calm area next to the shore, take time to examine some of the rocks in the shallows. If these nymphs are present you will notice them scurry about. Dark wingpads on the nymphs indicate they are close to emergence. If these shallow areas are left undisturbed, trout will move into them to feed on the emerging nymphs. This is challenging water to fish. Care must be taken on the approach and long leaders help to keep the fly line out of the trout's vision. When possible, fish down and across, using the shallow-nymph method, with the nymph or soft-hackle set below a strike indicator. When a strike indicator is not used, lightly weighted flies are needed to keep the fly off the bottom. As the fly lifts off the bottom at the end of the drift, twitch it gently.

Pale Evening Dun

Family: Heptageniidae
Genus: *Heptagenia*
Species: *solitaria*

Size: 7-10 mm
Underbody Color: Light olive-tan
Notice: Flattened head

J F M A M̲ J̲ J̲ A S O N D

Hatches of the Pale Evening Duns generally occur in the afternoon, early in the season, but will shift toward the evening hours in the warmer months. *Heptagenia* nymphs migrate out of moderate to fast riffles and runs into the slower waters along the margins of these areas. At emergence, the nymphs release from the river bottom and swim to the surface film. Once the nymphs reach the surface, the dun emerges swiftly. During the cooler days of the early season, or on stormy days, the adult flies tend to stay on the water for longer drifts. As the days warm later in the season, the duns normally fly off after a very short drift. When large numbers of *Heptagenia* hatch, trout will often feed selectively on either the nymphs or duns.

**Light Cahill, Comparadun, Olive-tan Sparkle Haystack
Hook Size: 12-16**

Fish will move into the calm, sometimes very shallow waters, next to riffles or runs and feed on these mayflies. If you notice flies on the water in these areas, stand back and take a few moments to observe the water. You may see a fish take a dun or see a subsurface boil from a trout feeding on a nymph. It is important not to rush in on this shallow, calm and clear water. Trout are very nervous in these waters and will bolt to deeper water at the slightest disturbance. Low water patterns, long fine tippets, a low profile and a low rod work best under these conditions. The Sparkle Haystack is a good emerger pattern to use when you aren't sure if the fish are feeding on duns or on nymphs in the surface film. Use a reach cast or any cast that will put the fly over the fish before the leader or fly line can alarm it.

March Brown Nymph

Family: Heptageniidae   **Size: 8-12 mm**
Genus: *Rhithrogena*   **Body Color: Olive-brown**
                                   Notice: Large gills

J F **M A** M J J A S O N D

With the help of unique gills that form a sucker-like disc, March Brown nymphs are able to cling like suction cups to the smooth rocks found in fast riffles. The grip they exert is evident the first time you try to remove one of these nymphs from a wet rock. This unique feature, along with the nymphs' flat body shape, helps keep them from being washed into the drift. These nymphs are poor swimmers and if sent tumbling in the current, they usually drift along until they reach the bottom. As the nymphs mature they migrate from fast to moderate currents in riffles. Emergence occurs in the afternoons, when the nymphs release their grip on the bottom and rise to the surface film, where they drift until the duns emerge. Trout feed heavily on these emerging nymphs.

**Gold-Ribbed Hare's Ear, March Brown Soft-Hackle,
March Brown Emerger
Hook Size: 10-14**

March Brown nymphs are available to trout when they migrate out of fast water into moderate flows. A Gold-Ribbed Hare's Ear fished dead-drift and close to the bottom, using the deep-nymph method, imitates these nymphs as they struggle to regain the bottom. Fish any seams where fast and slow currents meet. As duns start to appear in the afternoon, switch to the March Brown Soft-Hackle and use the deep-nymph method. But midway through the drift, slowly lift the fly toward the surface to imitate a rising nymph. Subsurface boils from feeding trout indicate the trout are feeding on nymphs close to the surface. Change to the shallow-nymph method, with the strike indicator three feet ahead of the fly. Fish down and across dead-drift, letting the fly drift naturally over the fish. At the end of the drift, allow the fly to swing to the surface. Trout will take the fly gently at any part of the drift. But as it starts to lift toward the surface they usually clobber it.

March Brown Dun

Family: Heptageniidae   **Size: 8-12 mm**
Genus: *Rhithrogena*   **Underbody Color: Light
                                                tan**

J F **M A** M J J A S O N D

Prior to emergence, the March Brown nymphs move from the fast water in riffles to moderate flows along the margins of riffles. Emergence occurs in the afternoon. The nymphs release their hold on the bottom to rise to the surface film, where the duns emerge from their nymphal shucks. During this emergence stage nymphs often drift a long way downstream before the duns emerge. It's not uncommon to see subsurface boils from trout feeding on emergers and floating nymphs a good distance upstream from the untouched drifting duns. On cooler days, duns often drift a long way before flying off to the shore. Spinner falls of egg-laying females occur in the evenings, upstream from riffles. Trout feed actively on these mayfly duns during the emergence, but the spinners don't ordinarily create a fishable spinner fall.

**March Brown Comparadun, Flick's March Brown,
Hairwing March Brown
Hook Size: 12-14**

Duns first appear during their afternoon emergence in water downstream from riffles. On a cool day, the duns may ride the water for long distances before flying off. Match the fly to the type of water on which the duns are floating. Use a high-floating fly, like Flick's March Brown, in choppy water. The Comparadun and Hairwing are low-riding flies well suited for calmer waters. It's important to recognize what stage of the March Brown the trout are taking. They often feed selectively on nymphs in the surface film during the first part of the hatch, switching to duns as the number of nymphs in the drift decreases. During and after the hatch, eddies downstream from riffles often have large numbers of crippled or trapped duns floating in the lazy-susan currents of the eddy, with trout selectively feeding on them. The long drift of the duns make this hatch a dry fly fisher's dream come true.

Slate-Winged Mahogany Dun

Family: Leptophlebiidae   **Size:** 7-9 mm
Genus: *Paraleptophlebia*   **Underbody Color:** Reddish-
J F M A M J J A S <u>O</u> N D   brown

Nymphs of these late-season mayflies live in fast flow-ing waters, moving to more moderate currents as they mature. Riffles, runs and flats with trailing weeds are ideal habitat for them. On emergence, the nymphs migrate to the margins of these waters, where the duns emerge. Mahogany duns normally spend a very long time on the water, some-times drifting 50 or 60 feet before taking wing. Generally, this is not a heavy hatch and is often sporadic throughout the day. The large amount of time the duns spend on the water makes them an attractive snack for trout. Fish typically take these duns from the surface of the water in slow, deliberate rises.

**Red Quill, Paraleptophlebia Comparadun, Mahogany Dun Thorax
Hook Size: 14-16**

Don't go searching for this hatch; it is too sporadic to reward the effort. But if you spend any time traveling the banks of the Deschutes River in the fall, you'll probably come across trout feeding on these mayflies. You will often notice the slow rise of a fish before you see any mayflies. Eddies are collecting areas for these long-floating Mahogany duns. Trout see them often enough during the hatch period to make a low-floating imitation a good searching pattern for eddies or edge water. Trout feeding on Mahogany duns in the slow, flat water on the margins of the river can provide some very challenging fishing. Use long fine leaders and a downstream or reach cast that puts the fly over the fish first. A little reconnaissance will help to put you in the best position for the careful presentation needed to hook fish in these difficult waters.

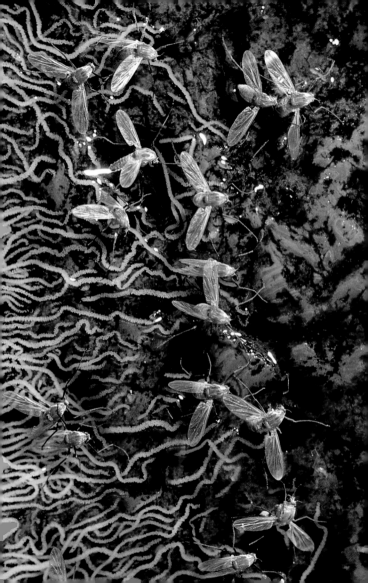

Chapter 8

Midges

Midges have complete metamorphosis. They go through four stages of development: egg, larva, pupa and adult. The Deschutes has midge hatches all through the year.

Midge larva are extremely abundant on the river bottom. Some are free-living, others burrow, and still others build cases. Midge larvae pupate in a larval case or in a burrow dug in silt on the river bottom.

Mature pupae float to the surface film and adults emerge from the pupal shuck. Adults can spend anywhere from a few seconds to a minute or more on the surface before flying off the water. Midge adults join in mating and egg-laying swarms over the water. Most midge females crawl under the water to lay their eggs.

Midge Pupa

Family: Chironomidae
J F M A M J J A S O N D

Size: 2-7 mm
Body Color: Black, dark brown

Midge larvae live in most of the waters of the Deschutes River. Larger concentrations of larvae are found in slow pools, where the larvae feed on debris that collect on the river bottom. Large numbers are also found in weedbeds. Midge larvae pupate on the bottom of the river. Once mature, pupae leave the bottom and swim with a slow wiggling motion to the surface. Midge pupae emerge all day long, throughout the year. Trout feed on the larvae, but the emerging pupae are more accessible and constantly in the drift. Even the largest trout will feed on these small pupae, particularly in the months with fewer hatches of other insects.

Serendipity, Krystal Flash Midge, TDC
Hook Size: 16-20

The best time to fish midge pupal imitations is between other insect hatches. That is, there is really no bad time to fish these imitations. The simplest way to fish these small flies is to use them as tailing flies behind larger nymphs. Tie one to the tippet above a larger fly as a dropper or fish one by itself with or without weight added to the leader. The two-fly system makes more sense on the Deschutes River. Why use split shot when a weighted stonefly nymph or a weighted nymph matching the current hatch will take the fly down and give trout a second choice? Use the short-line or deep-nymphing methods and fish inlets and tailouts of pools and around weed beds in runs and flats. Trout feeding on midge pupae in the surface film should be fished with an unweighted pupal imitation by itself or fished with the shallow-nymph method. Fish the pupa dead-drift, but it pays to lift the fly slowly once in awhile.

Midge Adult

Family: Chironomidae
J F M A M J J A S O N D

Size: 2-7 mm
**Body Color: Dark brown,
tan or black**

After leaving the river bottom and swimming to the surface, midge pupae drift along until the adult emerges from
the pupal shuck. Unable to completely release their pupal
shucks, a number of these emerging midges are stillborn, left
to float in the surface film. Midges emerge throughout the
year at any time of the day. In cooler months, midge adults
tend to remain on the water longer after emerging. Later,
midge adults fly out over the water in swarms to mate and
deposit their eggs. These large swarms often bring trout up
to feed on clusters of mating midge adults or individual egg-
laying females. Eddies and pools have large concentrations
of midge larvae. These areas also offer some of the best fishing to trout feeding on stillborn or adult midges.

**Stillborn CDC Midge, Black Midge, Griffith's Gnat
Hook Size: 18-22**

Trout feed on midge adults and any angler without a few adult midge patterns in his fly box must either trim down some other dry fly, in hopes of matching the hatch, or move on to other water. Trout feed on midge adults all year, but key on them more in winter and early spring when few other aquatic insects are available. In the flat waters of eddies and pools, trout need not expend much energy in feeding. Even a small midge becomes worth their effort, so watch for fish rising to midges in these areas. The small flies used to match midge adults are easier to see on flat water. Trout are normally wary in this type of water and any drag or careless cast may spook them. A long, fine leader helps promote a drag-free float and keeps the line out of the trout's feeding lane. Use the lower-riding flies in calm water. In faster water use the high-floating Griffith's Gnat, which also imitates clusters of midges on the surface.

Notice

You will notice that no weight has been add to the subsurface fly patterns. Weight should be added to any fly you want to reach the bottom quickly or you will need to add weight to the leader. It helps to have some subsurface fly patterns unweighted or lightly weighted when they are used as a dropper or a tailing fly. To keep track of the different weights use a different color thread on the fly heads final wrap for each weight.

A brass bead, available in varying sizes and weights, may be added behind the eye of a hook on the first step when tying a subsurface fly. It can replace any lead wrapped on the hook shank.

Other advantages are, the fly rides upright in the water like a natural insect and the brass bead adds sparkle to the fly.

FLY PATTERNS

GREEN ROCK WORM

Krystal Flash Green Rock Worm
Hook: TMC 3761,
   Mustad 3906B, sizes 12-16
Thread: Brown 6/0 prewaxed
Body: Peacock or Green
   Krystal Flash, twisted and
   wrapped
Head: Brown Haretron dubbing

Rhyacophila Caddis
Hook: TMC 200,
   Mustad 9671, sizes 12-16
Thread: Black 6/0 prewaxed
Rib: Fine green wire
Body: Light-green Haretron dubbing
Head: 75% Light-green
   Haretron and 25% Black
   Haretron dubbing

LaFontaine's Caddis Larva
Hook: TMC 200, sizes 12-16
Thread: Black 6/0 prewaxed
Rib: Stripped brown hackle
   stem
Body: Bright-green Antron dubbing
Thorax: Brown Antron dubbing
Legs: Brown grouse hackle
   fibers

GREEN SEDGE PUPA

Emergent Sparkle Pupa, Green
Hook: TMC 100,
   Mustad 94845, sizes 12-14
Thread: Black 6/0 prewaxed
Underbody: 25% olive and
   75% bright-green Antron dubbing
Overbody: Medium-olive
   Antron yarn
Legs: Grouse hackle fibers
Head: Brown dubbing

Krystal Flash Pupa
Hook: TMC 3761,
   Mustad 3609B, sizes 12-14
Thread: Brown 6/0 prewaxed
Underbody: Peacock or Green
   Krystal Flash, twisted and
   wrapped
Overbody: Light-tan Antron yarn
Head: Brown Haretron dubbing

Partridge and Green Soft-Hackle
Hook: TMC 3761,
   Mustad 3906, sizes 12-14
Thread: Olive 6/0 prewaxed
Body: Green silk floss

Thorax: Natural Haretron dubbing
Hackle: Gray partridge

GREEN SEDGE

Deer Hair Caddis, Olive
Hook: TMC 900BL,
   Mustad 94840, sizes 12-14
Thread: Tan 6/0 prewaxed
Body: Olive Haretron dubbing
Hackle: Dark blue dun hackle,
   palmered over body, clipped
   bottom
Wing: Natural dun deer hair

Diving Caddis, Green
Hook: TMC 3761,
   Mustad 3906B, sizes 12-14
Thread: Black 6/0 prewaxed
Body: Green Haretron dubbing
Underwing: Grouse hackle fibers
Overwing: Clear Antron fibers
Hackle: Brown neck hackle,
   sparse

Krystal Flash Diving Caddis
Hook: TMC 3761,
   Mustad 3906B, sizes 12-14
Thread: Brown 6/0 prewaxed
Tail: Clear Antron fibers
Body: Green Krystal Flash twist-
   ed and wrapped
Wing: Dark dun hen fibers

SPOTTED SEDGE LARVA

Dubbed Caddis Larva
Hook: TMC 3769,
   Mustad 3906B, sizes 12-14
Thread: Tan 6/0 prewaxed
Body: Green or Tan Haretron
   dubbing
Thorax : Brown Hareton dub-
bing, picked out

Latex Caddis
Hook: TMC 3761,
   Mustad 3906B, sizes 12-14
Thread: Brown 6/0 prewaxed
Underbody: Brown thread
Body: Latex strip, colored tan
   or green with waterproof
   marker
Thorax: Dark brown Haretron
   dubbing, picked out on bot-
   tom

Gold-Ribbed Hare's Ear
Hook: TMC 3769,
   Mustad 3906B, sizes 12-14
Thread: Black 6/0 prewaxed
Tail: Hare's mask guard hairs
Rib: Fine gold tinsel
Body: Hare's ear dubbing
Wing case: Mottled turkey
   quill
Thorax: Hare's ear dubbing

SPOTTED SEDGE PUPA

Deep Sparkle Pupa, Tan
Hook: TMC 100,
 Mustad 94845, sizes 12-14
Thread: Black 6/0 prewaxed
Underbody: Green Haretron
 dubbing
Overbody: Tan Antron yarn
Legs: Brown partridge hackle
Head: Brown Haretron dubbing

Emergent Sparkle Pupa, Tan
Hook: TMC 100,
 Mustad 94845, sizes 12-14
Thread: Brown 6/0 prewaxed
Underbody: Green Haretron
 dubbing
Overbody: Tan Antron yarn
Wing: Dark gray deer hair
Head: Brown Haretron dubbing

March Brown Spider Soft-Hackle
Hook: TMC 3761,
 Mustad 3906, sizes 12-14
Thread: Orange 6/0 prewaxed
Rib: Fine gold tinsel
Body: Natural Haretron dubbing
Hackle: Brown partridge

SPOTTED SEDGE

Deer Hair Caddis, Olive-Brown
Hook: TMC 900BL,
 Mustad 94840, sizes 12-14
Thread: Tan 6/0 prewaxed
Body: Olive-brown Haretron
 dubbing
Hackle: Dark blue dun hackle
Wing: Natural dun deer hair

Canoe Fly
Hook: TMC 900BL,
 Mustad 94840, sizes 12-14
Thread: Tan 6/0 prewaxed
Body: Olive-brown Haretron
 dubbing
Wing : Natural dun deer hair

Black Soft-Hackle
Hook: TMC 3761,
 Mustad 3906, sizes 12-14
Thread: Black 6/0 prewaxed
Body: Black Krystal Flash
 twisted and wrapped
Thorax: Brown Haretron dubbing
Wing: Dark dun hen hackle

LITTLE WESTERN WEEDY-WATER SEDGE LARVA

Herl Nymph
Hook: TMC 3761,
  Mustad 3906B, sizes 12-14
Thread: Black 6/0 prewaxed
Body: Peacock herl
Thorax: Black ostrich herl
Legs: Black hackle fibers

Drifting Cased Caddis
Hook: TMC 5262,
  Mustad 9671, sizes 12-14
Thread: Black 6/0 prewaxed
Body: Mottled turkey wing quill
  fibers, wrapped
Thorax: Green Antron dubbing
Legs: Black hackle fibers,
  stiff and clipped
Head: Black Antron dubbing

Peeking Caddis
Hook: TMC 3761,
  Mustad 3906B, sizes 12-14
Thread: Black 6/0 prewaxed
Rib: Fine oval gold tinsel
Body: Natural hare's ear
  dubbing
Legs: Ringneck pheasant back
  fibers
Head: Black ostrich herl

LITTLE WESTERN WEEDY-WATER SEDGE PUPA

Krystal Flash Pupa, Green
Hook: TMC 3761,
  Mustad 3906B, sizes 14-16
Thread: Brown 6/0 prewaxed
Underbody: Green Krystal
  Flash, twisted and wrapped
Overbody: Light tan Antron
  yarn
Head: Brown dubbing

Emergent Sparkle Pupa, Green
Hook: TMC 100,
  Mustad 94845, sizes 14-16
Thread: Black 6/0 prewaxed
Underbody: Green Antron
  dubbing
Overbody: Green Antron yarn
Wing: Dark gray deer hair
Head: Brown dubbing

CDC Caddis Emerger, Green
Hook: TMC 100,
  Mustad 94845, sizes 14-16
Thread: Olive 6/0 prewaxed
Tail: Olive Z-lon
Rib: Fine copper wire
Body: Green Antron dubbing
Antennae: Two woodduck
  flank fibers
Wing: Dark dun CDC feathers
Legs: Brown partridge hackle
Head: Dark brown Antron dub-
  bing

LITTLE WESTERN WEEDY-WATER SEDGE

Elk Hair Caddis, Olive
Hook: TMC 100,
   Mustad 94845, sizes 14-16
Thread: Olive 6/0 prewaxed
Rib: Fine gold wire
Body: Olive dubbing
Hackle: Medium blue dun,
   palmered over body
Wing: Light elk hair

Sparkle Caddis, Olive
Hook: TMC 100,
   Mustad 94845, sizes 14-16
Thread: Olive 6/0 prewaxed
Tail: Golden-olive Z-lon
Body: Olive poly dubbing
Wing: Natural deer hair

CDC Caddis, Olive
Hook : TMC 100,
   Mustad 94345, sizes 14-15
Thread : Olive 6/0 prewaxed
Body : Olive poly dubbing
Underwing : Light dun Z-lon
Wing : Medium dun CDC feather
Legs : Medium dun CDC feather
Thorax: Olive poly dubbing

LITTLE BLACK CADDIS LARVA

Gold-Ribbed Hare's Ear
Hook: TMC 2302,
   Mustad 3906B, sizes 14-18
Thread: Black 6/0 prewaxed
Tail: Hare's mask guard hairs
Rib: Fine gold tinsel
Body: Hare's ear dubbing
Wingcase: Mottled turkey quill
Thorax: Hare's ear dubbing

Cream Caddis Larva
Hook: TMC 3769,
   Mustad 3609, sizes 14-18
Thread: Brown 6/0 prewaxed
Body: Cream Haretron dubbing
Head: Dark brown Haretron
   dubbing

Latex Caddis
Hook: TMC 200,
   Mustad 3906B, sizes 14-18
Thread: Brown 6/0 prewaxed
Underbody: Brown thread
Body: Latex strip
Thorax: Dark brown dubbing,
   picked out on bottom

LITTLE BLACK CADDIS PUPA

Hare's Ear Soft-Hackle
Hook: TMC 3769,
   Mustad 3906, sizes 14-18
Thread: Tan 6/0 prewaxed
Body: Tan hare's ear dubbing
Hackle: Brown partridge

Little Green Caddis
Hook: Mustad 3906,
   sizes 14-18
Thread: Black 6/0 prewaxed
Body: Light-green Haretron
Legs: Blue dun hackle fibers
Wingcase: Gray mallard shoulder feather, trimmed
Head: Black ostrich herl

Deep Sparkle Pupa, Green
Hook: TMC 100,
   Mustad 94845, sizes 14-18
Thread: Black 6/0 prewaxed
Underbody: 30% olive Antron
   dubbing, 70% bright Antron
   dubbing
Overbody: Light-olive Antron
   yarn
Legs: Grouse hackle fibers
Head: Brown dubbing

LITTLE BLACK CADDIS

Deer Hair Caddis, Brown
Hook: TMC 900BL,
   Mustad 94845, sizes 14-18
Thread: Brown 6/0 prewaxed
Body: Dark brown dubbing
Hackle: Dark blue dun hackle,
   palmered over body, clip bottom
Wing: Natural dun deer hair

Starling and Herl
Hook: TMC 3769,
   Mustad 3906, sizes 14-18
Thread: Olive 6/0 prewaxed
Body: Peacock herl
Hackle: Starling wing cover
   feather

Krystal Flash Diving Caddis, Black
Hook: TMC 3769,
   Mustad 3906, sizes 14-18
Thread: Black 6/0 prewaxed
Tail: Clear Antron fibers
Body: Black Krystal Flash,
   twisted and wrapped
Wing: Dark dun hen hackle
   fibers

GIANT SALMONFLY NYMPH

Kaufmann's Black Stone
Hook: TMC 300,
  Mustad 9575, sizes 4-8
Thread: Black 6/0 prewaxed
Tail: Two black stripped goose
  fibers
Rib: Black Swannundaze
Body: Blend 50% black rabbit
  dubbing, 50% mixture of
  claret, amber, orange, rust,
  black, brown, blue, purple and
  ginger goat fur
Wingcase: Three separate sec-
  tions of lacquered dark turkey
  tail, clipped to shape
Thorax: Same as body
Head: Same as body
Antennae: Two black stripped
  goose fibers

Box Canyon Stone
Hook: TMC 3761,
  Mustad 3906B, sizes 2-6
Thread: Black 6/0 prewaxed
Tail: Dark brown stripped goose
  fibers
Body: Black yarn, twisted
Wingcase: Brown mottled
  turkey quill
Thorax: Black yarn
Hackle: Furnace hackle,
  wrapped over thorax

Lead-Eyed Woolly Bugger, Black
Hook: TMC 5263,
  Mustad 9672, sizes 6-8
Thread: Black 6/0 prewaxed
Tail: Black marabou
Body: Black chenille
Hackle: Black hackle, palmered
  over body
Eyes: Lead eyes

GIANT SALMONFLY

Bird's Stone
Hook: TMC 5212,
  Mustad 94831, sizes 4-8
Thread: Orange 6/0 prewaxed
Tail: Dark moose mane, two
  strands
Rib: Furnace saddle hackle,
  trimmed
Body: Orange floss
Wing: Natural brown bucktail
Hackle: Furnace saddle hackle,
  clipped top and bottom
Antennae: Dark moose mane,
  two strands

Improved Sofa Pillow
Hook: TMC 5263,
  Mustad 9672, sizes 4-8
Thread: Black 6/0 prewaxed
Tail: Natural elk hair
Rib: Brown hackle, undersized
Body: Orange wool yarn

Wing: Natural elk hair
Hackle: Brown saddle hackles

Clark's Stonefly
Hook: TMC 5263,
  Mustad 9672, sizes 4-8
Thread: Orange 6/0 prewaxed
Body: Flat gold tinsel
Underwing: Rust and gold
  macrame yarn, combed and
  mixed
Wing: Deer hair
Hackle: Brown saddle hackle

GOLDEN STONE NYMPH

Golden Stone Nymph
Hook: TMC 5263,
  Mustad 9672, sizes 6-8
Thread: Gold 6/0 prewaxed
Tail: Teal flank fibers
Rib: Gold thread
Shellback: Teal flank fibers
Body: Antique gold yarn
Legs: Teal flank fibers
Wingcase: Teal flank

Yellow Stone Nymph
Hook: TMC 5263,
  Mustad 9672, sizes 6-8
Thread: Brown 6/0 prewaxed
Tail: Cinnamon turkey quill,
  forked, three fibers per side
Rib: Antique gold yarn, one
  strand, and fine gold wire
Body: Yellow-brown yarn

Hackle: Grizzly and dyed
  brown grizzly hackle, wound
  over thorax
Gills: Light gray ostrich herl,
  wrapped with hackle

Kaufmann's Golden Stone
Hook: TMC 300,
  Mustad 9575, sizes 6-8
Thread: Gold 6/0 prewaxed
Tail: Two ginger stripped goose
  fibers
Rib: Pale ginger Swannundaze
Body: Blended, 50% golden
  brown rabbit dubbing, 50%
  mixture of claret, amber,
  orange, rust, black, brown,
  blue, purple and ginger goat
  fur
Wingcase: Three separate sections of lacquered mottled
  turkey wing quill, clipped to
  shape
Thorax: Same as body
Head: Same as body
Antennae: Two ginger stripped
  goose fibers

GOLDEN STONEFLY

Bucktail Caddis, Yellow
Hook: TMC 100,
  Mustad 94845, sizes 6-8
Thread: Yellow 6/0 prewaxed
Tail: Deer hair
Body: Yellow wool or poly yarn

Hackle: Brown hackle, palmered over body
Wing: Deer hair

Stimulator, Golden

Hook: TMC 200R, Mustad 94831 sizes 6-8
Thread: Fluorescent fire orange 6/0 prewaxed
Tail: Golden-brown elk hair
Rib: Fine gold wire and blue dun hackle
Body: Golden blend of goat (gold, ginger, amber, yellow) and golden brown Haretron
Wing: Golden-brown elk hair
Hackle: Furnace hackle, wrapped over the thorax
Thorax: Fluorescent fire orange Antron dubbing

Sofa Pillow

Hook: TMC 5212, Mustad 9671, sizes 6-8
Thread: Brown 6/0 prewaxed
Tail: Red goose quill sections
Body: Red floss, thin
Wing: Red fox squirrel tail
Hackle: Brown saddle hackle

LITTLE YELLOW STONEFLY NYMPH

Little Yellow Stone

Hook: TMC 300, Mustad 79580, sizes 10-12
Thread: Light-yellow 6/0 prewaxed
Tail: Dyed chartreuse mallard flank
Rib: Yellow thread
Shellback: Dyed chartreuse mallard flank
Body: Chartreuse wool yarn
Legs: Dyed chartreuse mallard flank fibers
Wingcase: Dyed chartreuse mallard flank fibers, one-third body length

Stonefly Creeper

Hook: TMC 5263, Mustad 9672, sizes 10-12
Thread: Yellow 6/0 prewaxed
Tail: Ringneck pheasant tail fibers
Shellback: Lemon woodduck flank fibers
Body: Stripped ginger hackle quill
Thorax: Amber goat fur dubbing
Legs: Brown partridge hackle, collar style

Maggot

Hook: TMC 5262, Mustad 9671, sizes 10-12
Thread: Black 6/0 prewaxed
Shellback: Ringneck pheasant tail fibers
Rib: Copper wire
Body: Pale yellow dubbing
Hackle: Brown hackle, sparse

LITTLE YELLOW STONEFLY

Bucktail Caddis, Yellow
Hook: TMC 100,
  Mustad 94845, sizes 10-12
Thread: Yellow 6/0 prewaxed
Tail: Deer hair
Body: Yellow wool or poly yarn
Hackle: Brown hackle,
  palmered over body
Wing: Deer hair

Stimulator, Yellow
Hook: TMC 200R, sizes 10-12
Thread: Yellow 6/0 prewaxed
Tail: Elk hair
Rib: Fine gold wire and brown
  hackle
Body: Yellow dubbing
Wing: Elk hair
Hackle: Brown hackle,wrapped
  over thorax
Thorax: Tan dubbing

Clark's Little Yellow Stone
Hook: TMC 5263,
  Mustad 9672, size 14
Thread: Yellow 6/0 prewaxed
Body: Flat gold tinsel
Underwing: Gold macrame
  yarn, combed
Wing: Deer hair
Hackle: Dyed yellow grizzly
  hackle

BLUE-WINGED OLIVE NYMPH

Baetis Nymph
Hook: TMC 200,
  Mustad 3906B, sizes 16-18
Thread: Olive 8/0 prewaxed
Tail: Dyed olive mallard flank
  fibers
Body: Medium olive dubbing
Wingcase: Black ostrich herl,
  short
Legs: Dyed olive mallard flank

Krystal Flash Nymph
Hook: TMC 200,
  Mustad 3906B, sizes 16-18
Thread: Black 8/0 prewaxed
Tail: Dark dun hen fibers
Body: Black Krystal Flash twist-
  ed and wrapped
Wingcase: Black Krystal Flash
  ends from body doubled over
  and left untwisted
Thorax: Olive brown dubbing,
  picked out on bottom

Baetis Soft-Hackle
Hook: TMC 3761,
  Mustad 9671, sizes 16-18
Thread: Gray 6/0 prewaxed
Tail: Blue dun hackle fibers
Body: Gray dubbing
Hackle: Blue dun hen hackle

BLUE-WINGED OLIVE

PALE MORNING DUN NYMPH

Little Olive
Hook: TMC 100,
   Mustad 94845, size 16-18
Thread: Olive 8/0 prewaxed
Tail: Blue dun hackle fibers
Body: Olive-brown dubbing
Wing: Blue dun hen hackle tips
Hackle: Blue dun hackle

Hairwing Dun
Hook: TMC 100,
   Mustad 94845, sizes 16-18
Thread: Olive 6/0 prewaxed
Tail: Dark blue dun hackle
   fibers, forked
Body: Olive dubbing
Hackle: Dark blue dun hackle,
   clipped on the bottom
Wing: Dark gray deer hair,
   butts to form head

Blue-Winged Olive Parachute
Hook: TMC 100,
   Mustad 94845, sizes 16-18
Thread: Olive 8/0 prewaxed
Tail: Dark blue dun hackle
   fibers
Body: Olive dubbing
Wing: Dun poly yarn
Hackle: Dark blue dun hackle,
   parachute style

Pale Morning Dun Nymph
Hook: Mustad 3906B,
   sizes 14-16
Thread: Olive 6/0 prewaxed
Tail: Dark brown partridge
Body: Olive-brown dubbing
Thorax: Dark brown dubbing
Legs: Dark brown partridge

Olive-Brown Hare's Ear
Hook: TMC 2302,
   Mustad 3906B, sizes 14-16
Thread: Olive 6/0 prewaxed
Tail: Olive-brown hare's mask
   guard hairs
Rid: Fine gold tinsel
Body: Olive-brown Haretron
Wingcase: Mottled turkey quill
Thorax: Olive-brown Haretron,
   picked out on bottom

PMD Floating Nymph
Hook: TMC 100,
   Mustad 94845, sizes 14-16
Thread: Olive 6/0 prewaxed
Tail: Ginger hackle fibers
Rib: Brown silk thread
Body: Olive-tan Antron dubbing
Wingcase: Dark dun poly dub-
   bing, shaped into a ball and
   placed on top of thorax area
Legs: Ginger hackle fibers

PALE MORNING DUN

CDC PMD Floating Nymph/Emerger
Hook: TMC 5230, Mustad 94833, sizes 14-16
Thread: Yellow 6/0 prewaxed
Tail: Woodduck flank fibers
Rib: Fine gold wire
Body: Yellow dubbing
Legs: Woodduck flank fibers
Wing: White CDC feathers, short
Thorax: Yellow dubbing

Pale Morning Comparadun
Hook: TMC 100, Mustad 94845, sizes 14-16
Thread: Olive 6/0 prewaxed
Tail: Medium gray dry hackle fibers, split
Body: Olive dubbing mixed with tinge of yellow
Wing: Light to medium deer hair

PMD Spinner
Hook: TMC 100, Mustad 94845, sizes 14-16
Thread: Brown 6/0 prewaxed
Tail: Bronze blue dun hackle fibers, split
Body: Reddish-brown with tinge of yellow, mixed
Wing: Bronze blue dun hackle, clipped on bottom

SLATE-WINGED OLIVE NYMPH

D D D
Hook: TMC 5263, Mustad 9672, sizes 10-14
Thread: Black 6/0 prewaxed
Tail: Ringneck pheasant tail fibers
Body: Peacock herl
Wingcase: Pheasant tail fibers
Thorax: Dark olive chenille
Legs: Pheasant tail fibers
Head: Peacock herl

Olive-Brown Hare's Ear
Hook: TMC 3761, Mustad 3906B, sizes 10-14
Thread: Brown 6/0 prewaxed
Tail: Dark olive-brown hare's ear guard fibers
Rib: Fine gold tinsel
Body: Dark olive-brown hare's ear dubbing
Wingcase: Dark turkey tail
Thorax: Dark olive-brown hare's ear dubbing

Green Drake Emerger
Hook: TMC 5262, Mustad 9671, sizes 10-14
Thread: Olive 6/0 prewaxed
Tail: Lemon woodduck flank fibers
Rib: Yellow silk thread
Body: Olive Haretron
Hackle: Dyed olive grizzly hen hackle

SLATE-WINGED OLIVE

Natural Dun (tied by Richard Bunse)
Hook: Mustad 94838, size 14
Thread: Black 6/0 prewaxed
Tail: Beaver fur guard hair fibers
Body: Packing foam, colored with a waterproof marking pen, olive-brown, extended
Wing: Natural gray dun deer hair

Western Green Drake
Hook: TMC 100,
   Mustad 94845, sizes 10-14
Thread: Olive 6/0 prewaxed
Tail: Dark moose body hair
Rib: Bright-olive floss, thin
Body: Olive dubbing
Wings: Dark moose body hair
Hackle: Grizzly hackle dyed olive

Olive Sparkle Haystack
Hook: TMC 100,
   Mustad 94845, sizes 10-14
Thread: Olive 6/0 prewaxed
Tail: Olive Antron
Rib: Yellow silk
Body: Olive Haretron
Wing: Dark deer hair, flared 180 degrees

PALE EVENING DUN NYMPH

Cate's Turkey
Hook: TMC 2302,
   Mustad 3906B, sizes 12-14
Thread: Black 6/0 prewaxed
Tail: Lemon woodduck flank fibers
Rib: Fine gold wire
Body: Turkey tail fibers, wrapped
Thorax: Peacock herl
Legs: Lemon woodduck flank fibers

Olive-brown Pheasant Tail
Hook: TMC 2302,
   Mustad 3906B, sizes 12-14
Thread: Brown 6/0 prewaxed
Tail: Dyed olive-brown ring-neck pheasant tail fibers
Rib: Fine copper wire
Body: Dyed olive-brown ring-neck pheasant tail fibers, wrapped
Wingcase: Dyed olive-brown ringneck pheasant tail fibers
Thorax: Peacock herl
Legs: Tips from wingcase
Head: Fine copper wire

Dark Olive Soft-Hackle
Hook: TMC 3769,
   Mustad 3906, sizes 12-14
Thread: Olive 6/0 prewaxed

Body: Olive-brown Antron dubbing
Hackle: Partridge

PALE EVENING DUN

Light Cahill
Hook: TMC 100,
  Mustad 94845, sizes 12-16
Thread: Cream 6/0 prewaxed
Tail: Light ginger hackle fibers
Body: Cream dubbing
Wings: Lemon woodduck flank
Hackle: Light ginger hackle

Comparadun
Hook: TMC 100,
  Mustad 94845, sizes 12-16
Thread: Cream 6/0 prewaxed
Tail: Light ginger hackle fibers,
  split
Body: Tan dubbing
Wing: Deer hair, flared 180
  degrees

Olive-tan Sparkle Haystack
Hook: TMC 100,
  Mustad 94845, sizes 12-16
Thread: Tan 6/0 prewaxed
Tail: Tan Antron
Body: Olive-tan dubbing
Wing: Deer hair, flared 180
  degrees

MARCH BROWN NYMPH

Gold-Ribbed Hare's Ear
Hook: TMC 2303,
  Mustad 3906B, sizes 10-14
Thread: Black 6/0 prewaxed
Tail: Hare's mask guard hairs
Rib: Fine oval gold tinsel
Body: Hare's ear dubbing
Wingcase: Mottled turkey quill
Thorax: Hare's ear dubbing

March Brown Soft-Hackle
Hook: Mustad 3906,
  sizes 10-14
Thread: Brown 6/0 prewaxed
Tail: Three pheasant tail fibers
Body: Dark brown hare's ear
Hackle: Brown partridge

March Brown Emerger
Hook: TMC 3769,
  Mustad 3906B, sizes 10-14
Thread: Brown 6/0 prewaxed
Tail: Three ringneck pheasant
  tail fibers
Rib: Gold thread, rib body only
Body: Hare's ear dubbing
Collar: Hare's ear dubbing,
  picked out
Wingcase: Brown partridge
  hackle fibers, short
Head: Hare's ear dubbiing

MARCH BROWN

March Brown Comparadun
Hook: TMC 100,
Mustad 94845, sizes 12-14
Thread: Tan 6/0 prewaxed
Tail: Beaver fur guard hairs or
Microfibetts, split
Body: Tan dubbing
Wing: Deer hair, flared 180
degrees

Flick's March Brown
Hook: TMC 100,
Mustad 94845, sizes 12-14
Thread: Tan 6/0 prewaxed
Tail: Dark ginger hackle fibers
Body: Tan dubbing
Wing: Woodduck flank fibers
Hackle: Brown and grizzly
hackle

Hairwing March Brown
Hook: TMC 100,
Mustad 94845, sizes 12-14
Thread: Tan 6/0 prewaxed
Tail: Light ginger hackle fibers
Body: Tan dubbing
Hackle: Blue dun hackle,
clipped on bottom
Wing: Deer hair, butts to form
head

SLATE-WINGED MAHOGANY DUN

Red Quill
Hook: TMC 100,
Mustad 94845, sizes 14-16
Thread: Gray 6/0 prewaxed
Tail: Medium bronze dun hackle fibers
Body: Coachman brown hackle
stem, stripped and wrapped
Wings: Lemon woodduck flank
fibers
Hackle: Medium bronze dun
hackle

**Paraleptophlebia
Comparadun**
Hook: TMC 100,
Mustad 94845, sizes 14-16
Thread: Black 6/0 prewaxed
Tails: Brown dry hackle fibers,
split
Body: Mahogany dubbing
Wing: Medium gray deer hair,
flared 180 degrees

Mahogany Dun Thorax
Hook: TMC 100,
Mustad 94845, sizes 14-16
Thread: Brown 6/0 prewaxed
Tail: Medium blue dun hackle
fibers, split
Body: Mahogany dubbing
Wing: Dark blue dun turkey
flats

Hackle: Medium blue dun
hackle, clipped on bottom

MIDGE PUPA

Serendipity
Hook: TMC 101,
Mustad 94859, sizes 16-20
Thread: Black 8/0 prewaxed
Body: Black dubbing
Head: Deer hair, cut short

Krystal Flash Midge
Hook: TMC 101,
Mustad 3906, sizes 16-20
Thread: Black 8/0 prewaxed
Body: Black Krystal Flash,
two strands twisted and
wrapped
Head: Dark brown dubbing

TDC
Hook: TMC 101,
Mustad 94859, sizes 16-18
Thread: Black 8/0 prewaxed
Rib: Fine silver tinsel,
wrapped over body
Body: Black dubbing
Thorax: Black dubbing
Head: White ostrich herl

MIDGE

Stillborn CDC Midge
Hook: TMC 101,
Mustad 94859, sizes 18-22
Thread: Black 8/0 prewaxed
Tail: Light dun CDC feather,
sparse
Body: Black dubbing
Wing: Light dun CDC feather

Black Midge
Hook: TMC 101,
Mustad 94859, sizes 18-22
Thread: Black 8/0 prewaxed
Tail: Black hackle fibers
Body: Black dubbing
Hackle: Black hackle

Griffith's Gnat
Hook: TMC 101,
Mustad 94859, sizes 18-22
Thread: Olive 8/0 prewaxed
Rib: Fine gold wire
Body : Peacock herl
Hackle: Grizzly hackle,
palmered over body

REFERENCES

Arbona, Fred, Jr., *Mayflies, the Angler, and the Trout.* Tulsa, Oklahoma: Winchester Press, 1980.

Edmonds, George, Jr., Steven Jensen and L. Berner, *The Mayflies of North and Central America.* Minneapolis, Minnesota: University of Minnesota Press, 1976.

Hafele, Richard, and Scott Roederer, *An Angler's Guide to Aquatic Insects and their Imitations.* Boulder, Colorado: Johnson Publishing Company, 1987.

Hafele, Rick, and Dave Hughes, *The Complete Book of Western Hatches.* Portland, Oregon: Frank Amato Publications, 1981.

LaFontaine, Gary, *Caddisflies.* New York: Nick Lyons Books, 1980.

Leiser, Eric, and Robert H. Boyle, *Stoneflies for the Angler.* New York: Alfred A. Knopf, 1982.

Richard, Carl, Doug Swisher and Fred Arbona, Jr., *Stoneflies.* New York: Nick Lyons Books/Winchester Press, 1980.

Salomon, Larry, and Eric Leiser, *The Caddis and the Angler.* Harrisburg, Pennsylvania: Stackpole Books, 1977.

INDEX